Oliver Marquardt

A plane-wave based k·p-model to study semiconductor nanostructures

Oliver Marquardt

A plane-wave based k·p-model to study semiconductor nanostructures

Studies of electronic properties of III-V nanostructures using continuum elasticity theory and an eight-band k·p-model

Südwestdeutscher Verlag für Hochschulschriften

Imprint

Any brand names and product names mentioned in this book are subject to trademark, brand or patent protection and are trademarks or registered trademarks of their respective holders. The use of brand names, product names, common names, trade names, product descriptions etc. even without a particular marking in this work is in no way to be construed to mean that such names may be regarded as unrestricted in respect of trademark and brand protection legislation and could thus be used by anyone.

Publisher:
Südwestdeutscher Verlag für Hochschulschriften
is a trademark of
Dodo Books Indian Ocean Ltd., member of the OmniScriptum S.R.L Publishing group
str. A.Russo 15, of. 61, Chisinau-2068, Republic of Moldova Europe
Printed at: see last page
ISBN: 978-3-8381-1935-9

Zugl. / Approved by: Paderborn, Universität Paderborn, Diss., 2010

Copyright © Oliver Marquardt
Copyright © 2010 Dodo Books Indian Ocean Ltd., member of the OmniScriptum S.R.L Publishing group

Contents

1 Introduction   **3**
- 1.1 Semiconductor nanostructures 3
  - 1.1.1 Design and modification of semiconductor nanostructures 4
  - 1.1.2 Theoretical modeling of semiconductor nanostructures 6
- 1.2 Continuum and atomistic models 6
- 1.3 Zero-, one- and two-dimensional nanostructures 7
- 1.4 Nanostructures grown in polar and nonpolar direction 8

2 Theoretical modeling   **9**
- 2.1 Electronic structure of semiconductors 9
- 2.2 The tight-binding method 13
- 2.3 The $\mathbf{k} \cdot \mathbf{p}$ method 16
  - 2.3.1 The basic $\mathbf{k} \cdot \mathbf{p}$ model without strain and built-in electric fields 16
  - 2.3.2 Deriving the $\mathbf{k} \cdot \mathbf{p}$ parameters from a given band structure 22
  - 2.3.3 Influence of strain and polarisation 23
  - 2.3.4 Calculation of strain field and polarisation potential 25
  - 2.3.5 Artificial symmetries in the continuum picture 28
- 2.4 The quantum confined Stark effect 30
- 2.5 Emission and absorption spectra 31
  - 2.5.1 Coulomb interaction 32
  - 2.5.2 Many-particle Hamiltonian 33

3 Implementation   **35**
- 3.1 The $\mathbf{k} \cdot \mathbf{p}$ formalism 35
- 3.2 Continuum elasticity theory 37
- 3.3 Conjugate-Gradient minimisation 38
  - 3.3.1 The basic Conjugate-Gradient algorithm 38
  - 3.3.2 The CG implementation for the eight band $\mathbf{k} \cdot \mathbf{p}$ formalism 40
  - 3.3.3 The CG implementation for the second-order continuum elasticity theory 41
  - 3.3.4 Preconditioning 41
  - 3.3.5 Time reversal symmetry 42
- 3.4 Piezoelectric potential 43

4	Applications		45
	4.1	Quantum dots	45
		4.1.1 Atomistic vs. continuum models	47
		4.1.2 Second vs. third-order elasticity	56
		4.1.3 Influence of size, shape and material composition on optical properties of GaN quantum dots	60
		4.1.4 Polar vs. nonpolar grown III-nitride quantum dots	65
	4.2	Quantum wires and dislocations	79
		4.2.1 Charge carrier localisation around screw dislocations in bulk GaN . .	80
		4.2.2 Wurtzite GaN quantum wires in vacuum	85
	4.3	Quantum wells and superlattices	89
		4.3.1 Polar InGaN/GaN superlattices	90
		4.3.2 Thickness fluctuations in nonpolar grown $In_{0.2}Ga_{0.8}N$ quantum wells .	98

5 Summary and outlook   **109**

A Wurtzite Hamiltonian $H_{8\times 8}$ **115**

B	Convergence tests and benchmarks	119
	B.1 Mesh accuracy	119
	B.2 Mesh softening	119
	B.3 Cell size	121
	B.4 Cutoff energy	121
	B.5 Time and memory consumption	123

C Fitting of k · p parameters   **127**

Chapter 1
Introduction

1.1 Semiconductor nanostructures

Semiconductor nanostructures have attracted much research interest within the past years due to their broad spectrum of possible applications ranging from novel light emission devices [86, 159, 196, 207] to single photon detectors and emitters and to applications in quantum computers [33, 205, 261]. The III-nitrides are of particular interest for optoelectronic applications, since their bulk band gaps in ternary alloys can in principle span the whole spectrum from infrared to ultraviolet light. For example, laser devices emitting blue or ultraviolet light are already based on III-nitride materials [167, 163, 229].

The decisive property of such devices is the localisation of charge carriers inside a nanostructure. In the context of this work, semiconductor nanostructures are structures with characteristic dimensions, e.g. height or base lengths, in the range of up to a few 100 nm and which, due to the bulk electronic properties of the involved semiconductor or isolator materials allow for a localisation of charge carriers inside or in the vicinity of the structure. This behaviour is commonly achieved by choosing chemical compounds with different electronic properties for the nanostructure and the surrounding material, which is called the **matrix material**.

More specifically, the localisation of charge carriers in semiconductor nanostructures is mainly determined by the conduction band (CB) and valence band (VB) edges of the involved materials (Fig. 1.1). If the conduction band minimum in the nanostructure is energetically lower than in the matrix material, electrons are expected to localise inside the nanostructure. Hole states localise inside the nanostructure, if the valence band maximum is higher there than in the surrounding matrix material. The main advantage of semiconductor nanostructures is the possibility to control the elastic, electronic and optical properties within a wide range according to the requirements of a specific application. Occuring quantum confinement effects, the binding energies as well as the localisation of charge carriers can be modified by varying shape and size of a nanostructure or the band edges of the nanostructure or the matrix material.

Semiconductor nanostructures exhibit length scales of a few up to some hundred nm

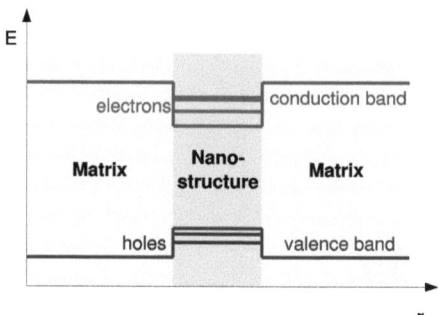

Figure 1.1: *Conduction (red) and valence band (blue) offset in a nanostructure and the surrounding matrix material. Additionally, the electron and hole localisation is depicted.*

and are commonly classified by the number of dimensions in which a localisation of charge carriers is realised. If charge carriers are free to move in two dimensions and are limited due to the band gap of the involved materials in one direction, we call these structures **quantum wells**. In **quantum wires**, a localisation of charge carriers is achieved in two dimensions, whereas free movement is possible in the third dimension. A localisation of charge carriers in all three dimensions is found in **quantum dots**. Correspondingly, strong quantum confinement effects can be seen below a certain size and charge carriers trapped in such a structure behave qualitatively similar as in atomic potentials. For the III-nitrides, characteristic dimensions of nanostructures which allow for significant quantisation effects are below 10 to 20 nm. Due to the occuring energy quantisation allowing discrete emission and absorption spectra, quantum dots are often referred to as **artificial atoms**.

1.1.1 Design and modification of semiconductor nanostructures

Light emission in semiconductor nanostructures requires coherent crystal interfaces between the nanostructure and the matrix, i.e. the nanostructure material adapts to the matrix lattice constants. In the epitaxial growth process (see Sec. 2.3.3) needed to ensure this coherent contact, thin layers of the nanostructure material are deposited on the surface of the underlying matrix material. During this process the material adapts the crystal structure and lattice constants of the matrix material, provided the difference between the lattice constants of the bulk matrix and nanostructure materials is sufficiently small, i.e. commonly below 5%, to allow for such an epitaxial growth. Common ways to perform such growth processes are the molecular beam epitaxy (MBE) [50] or the metal-organic chemical vapour decomposition (MOCVD) [149].

In the MBE, the elementary ingredients of a semiconductor (e.g., Ga and As or Al) are heated such that they sublimate in separate cells, in case they are not already in the gas phase. For the growth of III-nitrides, active nitrogen is produced in a radio-frequency plasma cell. The elementary ingredients then condensate at the substrate and react to form a regular cyrstalline structure. The term "beam" refers to the long mean free paths of the atoms due to the low pressure rather than to a directed and controlled beam of these atoms. By controlling the temperature and the amount of material evapourated, it is possible to control the growth rate of the process, i.e. the number of atomic monolayers grown in a certain time. Simultaneously, the material composition, e.g., when growing ternary alloys such as InGaN, can be controlled. The MBE allows a growth of high-quality crystals with a low number of defects and a small surface roughness. Moreover, the low growth rates allow for a well controlled thickness of films growing on a substrate using MBE. With MBE growth rates being typically in the order of only a few 100 nm/h [184] and the need for an ultra-high-vacuum, this method becomes rather costly and complicated.

For the MOCVD, sometimes also referred to as OMVPE (organometallic vapour phase epitaxy), metalorganic molecules are vapourised, which in principle allows much lower growth temperatures than the MBE and the growth process happens at moderate vapour pressures and does not require an UHV. Additionally, growth rates of a few ten μm/h can be achieved [127]. Despite the lower crystal quality and the higher surface roughness in comparison to MBE growth, the MOCVD is still the dominating growth procedure in industrial mass production of III-nitride light emitter devices.

The semiconductor layers at the substrate form by a chemical reaction. For the example of GaN grown on a substrate, Trimethylgallium (TMG), $(CH_3)_3Ga$ and NH_3 are used. The TMG molecule reacts at the surface and leaves the Ga as an adatom, which is incorporated in the surface. A similar process happens for the NH_3. Correspondingly, a GaN layer is formed and gaseous CH_4 forms as a by-product. The MOCVD commonly allows a faster growth process than the MBE, which is another advantage for industrial applications. However, due to the remaining vapour which is present during the growth process, a higher number of defects occurs, reducing the overall quality of the grown crystal.

The MBE and the MOCVD allow to control the growth velocity and the composition of the evapourated material by opening and shutting valves for the corresponding gases. This makes the growth of ternary and quaternary or alloys with even more components possible.

Therefore, direct control of the material composition and, thus, of the alloy band gaps and other material parameters, as well as layer thicknesses of semiconductor quantum wells is achieved in order to modify the electronic properties of such systems and therefore allows to desing crystals for the requirements of a specific application.

The growth process of quantum wires and dots is a more complex process and, therefore, not so easy to control. It is possible to etch quantum dots and wires with well defined shapes and sizes from bulk semiconductor crystals in lithographic processes. However, the dimensionality of such structures is limited by the wave length of the employed radiation and typical lateral dimensions for such structures are commonly in the order of 100 nm, which is too large to exhibit significant quantisation effects such as well defined discrete emission and absorption spectra. Correspondingly, these structures typically have the pure bulk material's electronic properties.

Self-assembled quantum dots begin to form when quantum wells grown on a substrate exceed a certain thickness (Stranski-Krastanov growth, see Sec. 2.3.3). The elastic energy stored in such systems due to the lattice constants of the layer material being larger than those of the matrix material makes the growth of three dimensional objects such as quantum dots or wires energetically favourable. The shape of such structures is mainly determined by the underlying crystal lattice and, therefore, not directly controllable via the parameters of the growth process. It is, however, possible to vary the average size of quantum dots in a limited range via the temperature during the growth process. Nevertheless, the dimensions of quantum dots in a quantum dot ensemble typically scatter in a certain range, i.e., it is difficult to grow quantum dots with a well-defined size, which makes the design of single, well defined dots for a specific application challenging [248].

Furthermore, it is possible to produce quantum dots in wet chemical fabrication processes [180], however, these systems are not subject of further discussion within this work.

1.1.2 Theoretical modeling of semiconductor nanostructures

Theoretical descriptions of semiconductor nanostructures are required to predict the properties of model quantum dot, wire and well systems. For example, it is possible to calculate the light emission spectrum of a given system and thus to design a nanostructure in a computational description such that it fits the requirements of a specific technical application, e.g. well-defined emission and absorption wavelengths for laser-based data storage systems. Furthermore, theoretical investigations can support experimental findings or explain effects which have been experimentally observed but lack a proper explanation. Last but not least, a computational theoretical model allows to neglect contributions of a given model in order to estimate their influence on the investigated system properties.

Employing a sufficiently detailed model, it is therefore possible to suggest parameters for shape, size and material composition of a nanostructure suited to the needs of a specific application. A theoretical modeling of such systems can thus save a lot of time, material and costs when performing systematic studies of a nanostructured system and its possible modifications.

1.2 Continuum and atomistic models

A number of theoretical approaches of different level of sophistication have been developed for the investigation of semiconductor nanostructures. These models range from analytic approaches [11, 252], numerical single band [99] effective mass models and more detailed multi-band $\mathbf{k} \cdot \mathbf{p}$ approaches [130, 71, 72, 221, 224] in a continuum picture to detailed atomistic descriptions such as the empirical tight-binding model (ETBM) [185, 210] or empirical pseudopotential calculations (EPM) [45, 244] and highly accurate ab initio calculations using the density functional theory (DFT) [106, 125].

While for some simple shaped nanostructures analytical methods were found to provide an excellent description [99], more complex nanostructure geometries, that are commonly observed in experiment cannot be treated analytically without a number of questionable simplifications and thus require more powerful numerical models. These can be classified as continuum and atomistic models.

In continuum single- or multi-band descriptions, the nanostructure and the matrix are modeled in an envelope function approach, i.e., material parameters are defined for the nanostructure and for the surrounding matrix, treating the whole system in a continuum picture. While the material parameters are commonly calculated using atomistic models, the nanostructure is modeled without taking single atomistic effects into account. Such continuum models are in principle not limited with respect to the system size, provided that the complexity of the nanostructure's geometry does not increase. On the other hand, it is clear that the possible impact of atomistic effects, e.g. in the vicinity of surfaces or defects, is completely neglected.

In atomistic approaches, single atoms of nanostructure and surrounding matrix are represented. This can be performed with different levels of sophistication, e.g. by taking interatomic interactions for a higher number of neighbouring atoms into account or considering more atomistic orbitals. It is clear, that the computational effort of such models generally increases with the number of involved atoms and, thus, with the system size, which limits these models in particular for studies of systematic modifications of a given reference nanostructure, where many similar calculations are made with only slight deviations of the system. Chapter 2 provides an introduction to atomistic ETBM and continuum $\mathbf{k} \cdot \mathbf{p}$ models.

For the purposes of the present work, a computationally cheap but still sufficiently accurate method which allows a high throughput of calculations to determine the electronic properties of a wide scale of III-nitride nanostructures is required, which, moreover, needs a highly efficient implementation. An eight band $\mathbf{k} \cdot \mathbf{p}$ model and a second-order continuum elasticity model were chosen to perform these studies. A novel plane-wave based implementation in an existing DFT program package allows the usage of highly optimised minimisation algorithms together with various advantages of a plane-wave formulation that enable us to perform all calculations with high efficiency. This formulation is discussed together with the details of the implementation in Chapter 3.

Correspondingly, one of the first objectives of the present work is to compare and evaluate different atomistic and continuum approaches to the electronic properties of semiconductor quantum dots in order to estimate possible errors which can be induced by a continuum description in the following studies and to verify the validity of the employed $\mathbf{k} \cdot \mathbf{p}$-model.

1.3 Zero-, one- and two-dimensional nanostructures

Within this work, an eight-band $\mathbf{k} \cdot \mathbf{p}$ model is applied together with a second-order continuum elasticity model to investigate a broad spectrum of III-nitride nanostructures. All of the studied systems are of particular interest for the design or improvement of novel light emitter devices. This work addresses different aspects of quantum dots, wires and wells such as the influence of size and shape or the material composition on the electronic properties. The resulting electronic states and binding energies are supposed to serve as input for many-particle calculations of the optical spectra of these nanostructures.

Besides conventional nanostructured systems where the nanostructure is a semiconductor material embedded in another matrix semiconductor, investigations on dislocations in bulk GaN are performed. Screw dislocations, which occur during the growth process in bulk GaN can possibly introduce a charge carrier localisation and thus act as unwanted radiative centers. In fact, experimental work confirms that such dislocations are non-radiative recombination centers [5, 59, 96, 104]. Within this work, the $\mathbf{k} \cdot \mathbf{p}$ formalism is used together with an analytical description of the shear strains resulting from the screw dislocation to study its influence on the charge carrier localisation. Chapter 4 provides in detail all results of the studies that have been performed within this work.

1.4 Nanostructures grown in polar and nonpolar direction

A special focus of this work is on the influence of the growth direction of a nanostructure on its electronic properties. III-nitride nanostructures commonly exhibit unusually strong polarisation potentials. Within the zincblende phase, these effects are often considered to be negligible, since due to the symmetry of the crystal structure rather small potentials are induced [254]. In the more stable wurtzite phase, however, much stronger polarisation potentials occur in typical nanostructures. In nanostructures grown along the polar [0001] crystal direction, these potentials are known to induce a strong spatial separation of electrons and holes and thus lead to poor radiative recombination rates and reduced oscillator strengths [34, 249]. These effects are expected to vanish completely in quantum well systems grown in a nonpolar direction and to be dramatically reduced in nonpolar grown quantum dot systems.

This work provides a detailed analysis of the origin of polarisation potentials and their influence on the charge carrier separation in different quantum well and quantum dot systems grown along polar and nonpolar directions. In particular, latest experimental informations concerning the structure of such systems are employed to provide a theoretical description of realistic quantum dot and quantum well systems. The results of these studies allow to make suggestions how to modify such nonpolar systems in order to achieve an optimal efficiency in light emitter applications.

Chapter 2

Theoretical modeling of optoelectronic properties in semiconductors

The aim of this work is to develop efficient methods to investigate the elastic and electronic properties of III-nitride nanostructures and to provide a detailed analysis of the electronic properties of various nanostructured systems. The electronic properties derived within this work can later on serve as input for the calculation of absorption spectra (for example performed by the semiconductor theory group (Prof. F. Jahnke) at the Institute for Theoretical Physics, University Bremen). The calculation of the absorption spectra of semiconductor nanostructures itself is therefore not subject of this work.

This chapter introduces the formalisms employed to obtain the electronic properties of semiconductor nanostructures. Special attention is paid to the analysis and comparison of atomistic and continuum descriptions of the electronic structure in bulk crystals to ensure the validity of the methods employed within this work. A detailed analysis of artificial symmetries occuring in the continuum eight band $\mathbf{k} \cdot \mathbf{p}$ model is given as well as a description of the quantum confined Stark effect. Furthermore, an introduction to second-order continuum elasticity theory is given, accounting for the influence of mechanical strain and polarisation potentials on the electronic structure.

2.1 Electronic structure of semiconductors

For the following considerations, we refer to the model of a solid in an ideal crystal structure. The ideal crystal is an infinite, regular lattice of a periodically repeated unit. The so-called **primitive cell**, is the smallest structure required to represent the whole crystal. To model the full, periodic crystal, lattice symmetry operations are required, which define the translation symmetry of the crystal. These **primitive translations** of the unit cell are given as:

$$\mathbf{R}_n = n_1 \mathbf{a_1} + n_2 \mathbf{a_2} + n_3 \mathbf{a_3}, \tag{2.1}$$

where the $\mathbf{a_i}$'s are the linearly independent basis vectors of the crystal and the n_i are integer numbers. The $\mathbf{R}_n$ form the **point lattice** of the crystal, i.e., the set of all $\mathbf{R}_n$'s leads to all equivalent points in the crystal [148].

With the knowledge of the primitive cell and the corresponding primitive translations, one can construct the **Wigner-Seitz cell**. For this purpose, one lattice point of the crystal is chosen as the center of the Wigner-Seitz cell. From this center, lines are drawn to all neighbouring lattice points. Perpendicular planes are then constructed in the middle of these connections. These planes are the boundaries of the Wigner-Seitz cell. Therefore, all points in this special primitive cell are closer to the cell center than to any center of a neighbouring cell.

The description of electronic wave functions is commonly performed in the **reciprocal space**. In this representation, a **k**-point lattice is defined similar to the point lattice in real space:

$$\mathbf{K}_m = m_1 \mathbf{b}_1 + m_2 \mathbf{b}_2 + m_3 \mathbf{b}_3, \tag{2.2}$$

where the m_i are again integers and the $\mathbf{b}_i$'s are the reciprocal basis vectors which are related to the real space basis vectors $\mathbf{a}_i$ by:

$$\mathbf{a}_i \cdot \mathbf{b}_j = 2\pi \delta_{ij}, \tag{2.3}$$

which means that $\mathbf{b}_i$ is perpendicular to $\mathbf{a}_k$ and $\mathbf{a}_j$ with $i,j,k = 1,2,3$ being cyclic. It therefore follows that:

$$\mathbf{b}_i = 2\pi \frac{\mathbf{a}_j \times \mathbf{a}_k}{\mathbf{a}_i \cdot (\mathbf{a}_j \times \mathbf{a}_k)} \quad \text{and} \quad \mathbf{a}_i = 2\pi \frac{\mathbf{b}_j \times \mathbf{b}_k}{\mathbf{b}_i \cdot (\mathbf{b}_j \times \mathbf{b}_k)}. \tag{2.4}$$

For each real space point lattice $\mathbf{R}_n$, a corresponding reciprocal space point lattice $\mathbf{K}_m$ exists and vice versa. Within the reciprocal lattice, a reciprocal unit cell can be constructed in similar manner to the Wigner-Seitz cell in real space. The Wigner-Seitz cell in reciprocal space is called the **Brillouin zone**.

The tool to calculate the electronic structure of a system is the Schrödinger equation. For the description of an electron in a crystal, we solve the one-particle Schrödinger equation:

$$\hat{H}|\Psi_{n,\mathbf{k}}(\mathbf{r})\rangle = \left(-\frac{\hbar^2}{2m}\nabla^2 + V(\mathbf{r})\right)|\Psi_{n,\mathbf{k}}(\mathbf{r})\rangle = \varepsilon_{n,\mathbf{k}}|\Psi_{n,\mathbf{k}}(\mathbf{r})\rangle, \tag{2.5}$$

with $|\Psi_{n,\mathbf{k}}\rangle$ being the nth electron wave function. The potential term $V(\mathbf{r}) = V(\mathbf{R}+\mathbf{r})$ already includes electron-electron interactions as well as the ionic potential. Solving the Schrödinger equation yields the electron wave function and eigenvalue of the nth state. The eigenvalues $\varepsilon_{n,\mathbf{k}} = \varepsilon_n(\mathbf{k})$ with **k** inside the Brillouin zone, describe the dispersion relation of the nth electron and are called **energy bands**. The complete function of all bands for the Brillouin zone, $\varepsilon_n(\mathbf{k})$, is referred to as the **band structure** of the crystal. An example band structure for Si is given in Fig. 2.1. In this plot, the band energies are given along different high symmetry paths inside the Brillouin zone. These paths are defined by so-called **high symmetry points**. For example, Γ denotes the center of the Brillouin zone

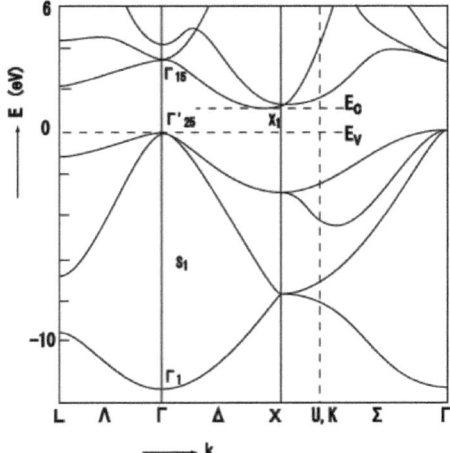

Figure 2.1: *Band energies as a function of the wave vector* **k** *for Si, taken from Ref. [44]. E_c denotes the bottom of the conduction band and E_v denotes the top of the valence band.*

($\mathbf{k} = (0,0,0)$) and X is the center of a squared face of the Brillouin zone in a face centered cubic lattice, $\mathbf{k} = (0, 2\pi/\mathbf{a}_2, 0)$. The band structure allows to calculate elementary properties of the solid, e.g. the band gap, which is the minimum difference between the lowest edge of the conduction band and the highest edge of the valence band, or the effective masses of electrons which can be obtained from the slope of the corresponding band.

There are different methods to calculate the band structure of an ideal crystal. Those methods which are of particular importance for the present work will be explained in more detail within the next sections.

The empirical pseudopotential method [45, 54, 69, 244] was developed in the 1960's and has proven that accurate semiconductor band structures can be obtained employing single electron models with rather simple potentials. In this method, atoms are represented by pseudopotentials and their shape is fitted to materials properties obtained from experiments or from first principles calculations.

Ab initio calculations based on density functional theory [106, 125] start with a many-particle problem which is reduced to the single particle Hamiltonian [70]

$$\left(-\sum_i^N \frac{\hbar^2}{2m_0} \nabla_i^2 + \sum_i^N V(\mathbf{r_1}) + V_{XC}(\mathbf{r}_i) + \sum_{i<j}^N U(\mathbf{r}_i, \mathbf{r}_j) \right) |\Psi\rangle = \varepsilon |\Psi\rangle, \qquad (2.6)$$

by capturing the many-particle potential $U(\mathbf{r}_i, \mathbf{r}_j)$ in terms of a particle density which contains all electron-electron Coulomb repulsions, allowing a massive reduction of the computational effort. $V_{XC}(\mathbf{r}_i)$ contains all many-particle interactions and is called the exchange-correlation potential. ε is the single-particle energy. Except for the case of a free electron gas, the exchange-correlation functionals are not known. Therefore, different approximations are typically made. However, both commonly used approximations, the local-density approximation (LDA) [42] as well as the generalised gradient approximation (GGA) [181] typically have difficulties to correctly describe some fundamental material properties in semiconducting and isolating systems such as, e.g., the band gap. In the case of the semiconducting InN, e.g., these effects even lead to predicting the material to be metallic. For such systems, the lack of knowledge of the correct exchange-correlation functional can be compensated by applying many-body perturbation theory based on the so-called Green's function [174]. For example, combining DFT with the many-particle G_0W_0-method [101], leads to an excellent description of experimentally accessible material properties related to the band structure for sp-valent systems such as InN, GaN and AlN [10, 176, 201, 202]. Such highly accurate ab initio calculations allow the calculation of input parameters for models with a lower level of sophistication. Due to their huge computational costs, however, they are limited to a small number of atoms per unit cell and are thus not well suited to describe nanostructured systems which can typically contain 10^5 to 10^7 atoms directly.

Atomistic tight-binding approaches allow the treatment of a larger number of atoms, in particular these approaches are able to handle the large number of atoms which have to be considered in common III-nitride nanostructures. Tight-binding models calculate wave functions using a basis set which consists of a limited number of atomic orbitals involved in binding processes. In most cases, interactions are limited to the atomic orbitals of second or third nearest neighbours. As will be explained in more detail within the next section, the electronic dispersion is obtained as a function of tight-binding parameters which are fitted such that they reproduce the band structure and can thus provide the physically meaningful Kohn-Luttinger parameters, effective electron masses, band gap and spin-orbit and crystal-field splittings at the Γ-point [94]. Another fitting of the tight-binding parameters around other high-symmetry points is required in order to reproduce an accurate description of the band structure throughout the rest of the Brillouin zone.

However, for the simulation of electronic properties of devices based on nanostructured systems like quantum dots, wires and wells, the important region of the band structure in direct semiconductors is the Γ-point due to the dimensions of the structure typically being in the range of a few nm. Limiting oneself to small $\mathbf{k}$-vectors one can use the $\mathbf{k} \cdot \mathbf{p}$ perturbation theory [13, 21, 87]. This chapter provides an introduction into the well-established $\mathbf{k} \cdot \mathbf{p}$-formalism including strain and polarisation effects. Additionally, the continuum elasticity model applied within this work to account for strain and piezoelectric effects in epitaxially grown, lattice mismatched nanostructures is explained. An introduction to the atomistic empirical tight-binding method (ETBM) and its effective bond-orbital model (EBOM) simplification is given as well in order to provide the theoretical background for the comparison of atomistic and continuum models given in Sec. 4.1.1.

2.2 The tight-binding method

Within the ETBM, the time-independent Schrödinger equation

$$\hat{H}|\Psi_{n,\mathbf{k}}\rangle = \varepsilon_n(\mathbf{k})|\Psi_{n,\mathbf{k}}\rangle \quad \text{with} \quad \hat{H} = \frac{\hat{\mathbf{p}}^2}{2m_0} + V(\mathbf{r}), \tag{2.7}$$

where $\hat{\mathbf{p}}$ is the momentum operator and $V(\mathbf{r})$ is the lattice-periodic potential, is solved by expanding $|\Psi_{n,\mathbf{k}}\rangle$ into a linear combination of atomic orbitals $|\Phi_\alpha(\mathbf{r}-\mathbf{R})\rangle$ centered at the atom positions $\mathbf{R}$ of the underlying crystal. With N being the number of atoms in a given macroscopic volume, so-called **Bloch sums** are constructed, linear combinations of atomic orbitals of the form

$$|\Psi_{n,\mathbf{k}}(\mathbf{r}+\mathbf{R})\rangle = e^{i\mathbf{k}\mathbf{R}}|\Psi_{n,\mathbf{k}}(\mathbf{r})\rangle \tag{2.8}$$

that satisfy the Bloch condition (Bloch theorem). These Bloch sums are given by the discrete Fourier transforms of the atomic orbitals:

$$|\chi_{\alpha\mathbf{k}}(\mathbf{r})\rangle = \frac{1}{\sqrt{N}}\sum_\mathbf{R} e^{i\mathbf{k}\mathbf{R}}|\Phi_\alpha(\mathbf{r}-\mathbf{R})\rangle \tag{2.9}$$

where $\mathbf{k}$ is the wave vector and $\mathbf{R}$ runs over all N lattice sites. Following Ref. [144], the solutions of the Schrödinger equation $|\Psi_{n\mathbf{k}}\rangle$ are expanded in terms of the Bloch sums as

$$|\Psi_{n,\mathbf{k}}\rangle = \sum_{\alpha'\mathbf{k}'} c_{\alpha'}(\mathbf{k}')|\chi_{\alpha',n,\mathbf{k}'}\rangle, \tag{2.10}$$

with the tight-binding coefficients $c_{\alpha'}$. With the matrix elements

$$\langle\chi_{\alpha\mathbf{k}}|\hat{H}|\chi_{\alpha'\mathbf{k}'}\rangle = \frac{1}{N}\sum_{\mathbf{R}\mathbf{R}'} e^{-i\mathbf{k}\mathbf{R}} e^{i\mathbf{k}'\mathbf{R}'} \int d\mathbf{r} \langle\Phi_\alpha(\mathbf{r}-\mathbf{R})|\hat{H}(\mathbf{r})|\Phi_{\alpha'}(\mathbf{r}-\mathbf{R}')\rangle, \tag{2.11}$$

one can now determine the band energies in Eq. (2.7) by solving a reduced matrix equation:

$$\sum_{\alpha'} \langle\chi_{\alpha\mathbf{k}}|\hat{H}|\chi_{\alpha'\mathbf{k}}\rangle c_{\alpha'}(\mathbf{k}) = \varepsilon(\mathbf{k}) c_\alpha(\mathbf{k}). \tag{2.12}$$

Besides an approximation that the $\chi_{\alpha,\mathbf{k}}$'s are orthonormal, Eq. (2.12) is simply a reformulation of the Schrödinger equation (2.7). The advantage of this formulation is the direct access to the so-called **tight-binding approximation**. The representation of charge carriers based on atomic orbitals reflects the case that the electrons are tightly bound to the atoms if interatomic distances are large. However, even for interatomic distances of realistic crystals, a construction of the $\chi_{\alpha\mathbf{k}}$'s from only a few atomic orbitals Φ_α in Eq. (2.9) was found to provide an excellent approximation for actual band structure calculations [144]. As a consequence, the construction of $|\Psi_{n,\mathbf{k}}\rangle$ is also restricted to a finite number of Bloch sums in Eq. (2.10). Furthermore, the restriction to a reasonable number of nearest neighbours (commonly nearest and next-nearest neighbours) reduces the summation in Eq. (2.11) and thus allows to diagonalise a finite-dimensional matrix in Eq. (2.12) in order to calculate wave functions and energies. The exact form of such a matrix depends on the underlying crystal lattice, the shape of the atomic orbitals and the number of involved nearest neighbours. Within a realistic crystal, complicated atomic potentials and interactions inbetween the electrons require to fit a large number of parameters for the matrix elements $\langle \chi_{\alpha,\mathbf{k}}|\hat{H}|\chi_{\alpha'\mathbf{k}'}\rangle$ in Eq. (2.12) to experimentally determined or ab initio calculated band structures, making this model complicated and cumbersome [144]. Taking more nearest neighbours or more atomic orbitals into account allows a better description of the band structure even for indirect band gap semiconductors but correspondingly raises the computational effort.

The **effective bond-orbital model** [43] is a simplification of the ETBM that needs a much smaller parameter set which can be directly related to the physically meaningful $\mathbf{k}\cdot\mathbf{p}$ parameters. This model is computationally cheaper than the ETBM and it is therefore worth to be discussed when choosing a suitable model for the investigations performed within this work. Employing the s orbitals of the cations and the p_x, p_y and p_z orbitals of the anions, the EBOM provides a tight-binding matrix that reproduces exactly the eight band $\mathbf{k}\cdot\mathbf{p}$ Hamiltonian around the Γ-point (see Sec. 2.3.1) [143]. Within the EBOM, a wave function $|\Psi_{n,\mathbf{k}}\rangle$ is constructed from the atomic orbitals

$$\Phi_s(\mathbf{r}) = R_s(r)s(\theta,\varphi), \quad \Phi_{p_i} = R_p(r)p_i(\theta,\varphi) \quad i=x,y,z.$$

The radial contributions $R_s(r)$ and $R_p(r)$ are unimportant for further considerations, since we do not calculate any of the integrals in Eq. (2.11) and we furthermore assume that the radial contributions vanish rapidly at infinity. For the example of a zincblende lattice, these effective orbitals are associated with an underlying fcc-lattice which artificially adds inversion symmetry that is actually not given due to the anions and cations being different atoms. Following Loehr [144], the Hamiltonian matrix equation Eq. (2.12) can now be written as:

$$\begin{pmatrix} \langle\chi_s|\hat{H}|\chi_s\rangle & \langle\chi_s|\hat{H}|\chi_{p_x}\rangle & \langle\chi_s|\hat{H}|\chi_{p_y}\rangle & \langle\chi_s|\hat{H}|\chi_{p_z}\rangle \\ \langle\chi_{p_x}|\hat{H}|\chi_s\rangle & \langle\chi_{p_x}|\hat{H}|\chi_{p_x}\rangle & \langle\chi_{p_x}|\hat{H}|\chi_{p_y}\rangle & \langle\chi_{p_x}|\hat{H}|\chi_{p_z}\rangle \\ \langle\chi_{p_y}|\hat{H}|\chi_s\rangle & \langle\chi_{p_y}|\hat{H}|\chi_{p_x}\rangle & \langle\chi_{p_y}|\hat{H}|\chi_{p_y}\rangle & \langle\chi_{p_y}|\hat{H}|\chi_{p_z}\rangle \\ \langle\chi_{p_z}|\hat{H}|\chi_s\rangle & \langle\chi_{p_z}|\hat{H}|\chi_{p_x}\rangle & \langle\chi_{p_z}|\hat{H}|\chi_{p_y}\rangle & \langle\chi_{p_z}|\hat{H}|\chi_{p_z}\rangle \end{pmatrix} \begin{pmatrix} c_s(\mathbf{k}) \\ c_{p_x}(\mathbf{k}) \\ c_{p_y}(\mathbf{k}) \\ c_{p_z}(\mathbf{k}) \end{pmatrix} = \varepsilon(\mathbf{k}) \begin{pmatrix} c_s(\mathbf{k}) \\ c_{p_x}(\mathbf{k}) \\ c_{p_y}(\mathbf{k}) \\ c_{p_z}(\mathbf{k}) \end{pmatrix}.$$
(2.13)

The single matrix elements are now calculated via

$$\langle\chi_\alpha|\hat{H}|\chi_\beta\rangle = \sum_{\mathbf{R}\in\{\mathbf{R}_n\}} e^{i\mathbf{k}\mathbf{R}} \int d\mathbf{r}\langle\Phi_\alpha(\mathbf{r})|\hat{H}(\mathbf{r})|\Phi_\beta(\mathbf{r}-\mathbf{R})\rangle, \tag{2.14}$$

where $\{\mathbf{R}_n\}$ are the positions of the neighbours taken into account (typically nearest and next nearest neighbours). The integral in Eq. (2.14) can be expressed as a real number for each neighbouring atom:

$$\varepsilon_{\alpha\beta}(\mathbf{R}) = \int d\mathbf{r}\langle\Phi_\alpha(\mathbf{r})|\hat{H}(\mathbf{r})|\Phi_\beta(\mathbf{r}-\mathbf{R})\rangle. \tag{2.15}$$

This allows to write Eq. (2.14) as

$$\langle\chi_\alpha|\hat{H}|\chi_\beta\rangle = \sum_{\mathbf{R}\in\{\mathbf{R}_n\}} e^{i\mathbf{k}\mathbf{R}} \varepsilon_{\alpha\beta}(\mathbf{R}). \tag{2.16}$$

For the example of an fcc lattice, the Hamiltonian in Eq. (2.13) can be approximated by a Taylor expansion around $\mathbf{k} = \mathbf{0}$ to second order in $\mathbf{k} = (k_x, k_y, k_z)$ with E_{cb} being the conduction band minimum and E_{vb} being the valence band maximum:

$$\hat{H} = \begin{pmatrix} E_{\text{cb}} & 0 & 0 & 0 \\ 0 & E_{\text{vb}} & 0 & 0 \\ 0 & 0 & E_{\text{vb}} & 0 \\ 0 & 0 & 0 & E_{\text{vb}} \end{pmatrix} \quad (2.17)$$

$$+ \begin{pmatrix} D'\mathbf{k}^2 & i\nu k_x & i\nu k_y & i\nu k_z \\ -i\nu k_x & -A'k_x^2 - B'(k_y^2 + k_z^2) & -C'k_x k_y & -C'k_x k_z \\ -i\nu k_y & -C'k_x k_y & -A'k_y^2 - B'(k_x^2 + k_z^2) & -C'k_y k_z \\ -i\nu k_z & -C'k_x k_z & -C'k_y k_z & -A'k_z^2 - B'(k_x^2 + k_y^2) \end{pmatrix}$$

where nearest and next nearest neighbour interactions are taken into account for a basis set restricted to the s, p_x, p_y and p_z orbitals in the parameters A', B', C', D' and ν:

$$\begin{aligned} A' &= a^2(E_{xx}(1,1,0) + 2E_{xx}(2,0,0)) \\ B' &= a^2(E_{xx}(1,1,0) + 1/2 E_{xy}(1,1,0) + 2E_{xx}(0,0,2)) \\ C' &= a^2 \frac{1}{2}(E_{xx}(1,1,0) + E_{xy}(1,1,0) - E_{xx}(0,1,1)) \\ D' &= -a^2(E_{ss}(1,1,0) + E_{ss}(2,0,0)) \\ \nu &= 2\sqrt{2}a E_{sx}(1,0,0), \end{aligned}$$

with a being the lattice constant. The $E_{xx}(i,j,k)$ and $E_{sx}(i,j,k)$ overlap elements are equal to the corresponding elements for the y and z direction in an fcc lattice. Following Loehr [144], these parameters can now be related to the physically meaningful Luttinger parameters γ_i and the effective electron mass m_e using the relation

$$A = A' + \frac{\nu^2}{E_g}, \quad B = B', \quad C = C' + \frac{\nu^2}{E_g}, \quad D = D' + \frac{\nu^2}{E_g} \quad (2.18)$$

and

$$\frac{\hbar^2}{2m_0}\gamma_1 = \frac{A + 2B}{3}, \quad \frac{\hbar^2}{2m_0}\gamma_2 = \frac{A - B}{6}, \quad \frac{\hbar^2}{2m_0}\gamma_3 = \frac{C}{6}, \quad \frac{\hbar^2}{2m_0 m_e} = D \quad (2.19)$$

where $E_g = E_{\text{cb}} - E_{\text{vb}}$ is the band gap. These parameters provide a direct connection between the atomistic EBOM model and the continuum $\mathbf{k} \cdot \mathbf{p}$-model (see Sec. 2.3). The relation between the Luttinger parameters and the valence band effective masses will be given in the next section. The EBOM can excellently reproduce the band structure throughout the whole Brillouin zone by solving Eq. (2.13) for different $\mathbf{k}$-points, provided a sufficiently large number of basic atomic orbitals and neighbour interactions is taken into account. However, the atomistic nature of the EBOM implies a computational effort growing linearly with the system size.

2.3 The k·p method

The computational effort of atomistic models such as the ETBM and the EBOM always depends on the number of involved atoms and thus on the system size and complexity. Continuum models like the effective mass approximation (EMA) [214, 252] or **k · p** models of different levels of sophistication do not describe the atomistic nature of the system and, therefore, do not require higher computational costs for larger systems, as long as the complexity of the structure does not increase. This makes such models the ideal tool for the investigation of electronic properties of semiconductor nanostructures with characteristic dimensions in the order of a few ten nm and containing millions of atoms.

2.3.1 The basic k · p model without strain and built-in electric fields

The **k · p** theory is a perturbative approach for solving the one-particle Schrödinger equation [115, 116]. It is usually combined with the envelope potential method [20, 38], where a wave function in a coherent crystal is expressed by a product of a Bloch term oscillating with the unit cell length and a macroscopic envelope potential contribution [38]:

$$\Psi(\mathbf{r}) = \sum_n F_n(\mathbf{r}) U_n(\mathbf{r}), \tag{2.20}$$

where the $U_n(\mathbf{r})$ are the rapidly oscillating Bloch wave functions and $F_n(\mathbf{r})$ is the slowly varying envelope function. A prerequisite for the application of the **k · p** formalism is that this macroscopic potential varies slowly compared to the lattice periodic atomic potential. If this is the case, a description of nanostructured semiconductor systems like quantum wells, wires and dots employing the envelope potential method becomes possible by treating the involved material parameters as spatially dependent [37, 74].

The **k · p** model can be derived by substituting Bloch wave functions (2.8) into the one-particle Schrödinger equation:

$$\left[\frac{\hat{\mathbf{p}}^2}{2m_0} + V(\mathbf{r})\right] |\Psi_{n,\mathbf{k}}(\mathbf{r})\rangle = \varepsilon_n(\mathbf{k}) |\Psi_{n,\mathbf{k}}(\mathbf{r})\rangle. \tag{2.21}$$

With the atomic potential term being lattice periodic, $V(\mathbf{r}) = V(\mathbf{r} + \mathbf{R})$ and employing Bloch wave functions

$$|\Psi_{n,\mathbf{k}}(\mathbf{r})\rangle = e^{i\mathbf{k}\mathbf{r}} |u_{n,\mathbf{k}}(\mathbf{r})\rangle, \tag{2.22}$$

one finds the **k · p** equation [256]:

$$\left[\frac{\hat{\mathbf{p}}^2}{2m_0} + \frac{\hbar \mathbf{k} \cdot \hat{\mathbf{p}}}{m_0} + \frac{\hbar^2 k^2}{2m_0} + V(\mathbf{r})\right] |u_{n,\mathbf{k}}(\mathbf{r})\rangle = \varepsilon_n(\mathbf{k}) |u_{n,\mathbf{k}}(\mathbf{r})\rangle. \tag{2.23}$$

The occurence of a scalar product **k·p** in Eq. (2.23) is the reason for calling the here described method the **k · p**-formalism. The perturbation is commonly performed at the Brillouin zone center $\mathbf{k}_0 = (0,0,0)$, for which Eq. (2.23) reduces to

$$\left[\frac{\hat{\mathbf{p}}^2}{2m_0} + V(\mathbf{r})\right] |u_{n,0}(\mathbf{r})\rangle = \varepsilon_n(\mathbf{0})|u_{n,0}(\mathbf{r})\rangle. \tag{2.24}$$

However, similar equations can be derived for any other $\mathbf{k} = \mathbf{k}_0$ within the Brillouin zone [256]. Once $\varepsilon_n(\mathbf{0})$ and $|u_{n,0}(\mathbf{r})\rangle$ are known, the terms linear and quadratic in $\mathbf{k}$ in Eq. (2.23) can be treated as small deviations. The Hamiltonian in Eq. (2.24) has the symmetry properties of the crystal point group provided by the potential term V, since the kinetic contribution $\mathbf{p}^2/2m_0$ has rotational symmetry.

Using the notation of Yu and Cardona [256], a $\mathbf{k}\cdot\mathbf{p}$-Hamiltonian treating only a certain number of bands (e.g., the lowest conduction band and the three highest valence bands) exactly (this set is labeled Γ_e) and all other bands perturbatively can be written as follows:

$$\hat{H}_{ij} = \varepsilon_i \delta_{ij} + \hat{H}_{ij}^0 + \sum_{l \notin \Gamma_e} \frac{\hat{H}_{il}\hat{H}_{lj}}{\varepsilon_i - \varepsilon_l} \tag{2.25}$$

$$= \varepsilon_i \delta_{ij} + \left\langle \Psi_i \left| \frac{\hbar^2 \mathbf{k}^2}{2m_0} + \frac{\hbar \mathbf{k}\cdot\mathbf{p}}{m_0} \right| \Psi_j \right\rangle$$

$$+ \sum_{l \notin \Gamma_e} \left\langle \Psi_i \left| \frac{\hbar^2 \mathbf{k}^2}{2m_0} + \frac{\hbar \mathbf{k}\cdot\mathbf{p}}{m_0} \right| \Psi_l \right\rangle \left\langle \Psi_l \left| \frac{\hbar^2 \mathbf{k}^2}{2m_0} + \frac{\hbar \mathbf{k}\cdot\mathbf{p}}{m_0} \right| \Psi_j \right\rangle \frac{1}{\varepsilon_i - \varepsilon_l}. \tag{2.26}$$

This allows in principle the formulation of $\mathbf{k}\cdot\mathbf{p}$-models of different levels of sophistication with an accuracy depending on the number of bands included in Γ_e. As an example, the element H_{11} of a 4×4 Hamiltonian neglecting the spin-orbit splitting shall be derived here for the zincblende crystal structure, consistent with the above description of the EBOM. For this case, basis functions Ψ_i in Eq. (2.26) are the $|\Psi_s\rangle$, $|\Psi_{p_x}\rangle$, $|\Psi_{p_y}\rangle$ and $|\Psi_{p_z}\rangle$-bands, consistent with the considerations for the EBOM model in Sec. 2.2. The Hamiltonian element $\hat{H}_{11} = \hat{H}_{ss}$ can, according to Eq. (2.26), be written as:

$$\hat{H}_{11} = E_{\text{cb}} + \frac{\hbar^2 \mathbf{k}^2}{2m_0} + \langle \Psi_s | \frac{\hbar \mathbf{k}\cdot\mathbf{p}}{m_0} | \Psi_s \rangle$$

$$+ \sum_{l \neq s, p_x, p_y, p_z} \langle \Psi_s | \frac{\hbar \mathbf{k}\cdot\mathbf{p}}{m_0} | \Psi_l \rangle \langle \Psi_l | \frac{\hbar \mathbf{k}\cdot\mathbf{p}}{m_0} | \Psi_s \rangle \frac{1}{E_{\text{cb}} - \varepsilon_l} \tag{2.27}$$

The $|\Psi_s\rangle$-band corresponds to the lowest lying conduction band, $\varepsilon_s \delta_{ss} = E_{\text{cb}}$ therefore is the conduction band edge. The upper valence bands $|\Psi_{p_x}\rangle$, $|\Psi_{p_y}\rangle$ and $|\Psi_{p_z}\rangle$ are energetically degenerate at $\mathbf{k} = \mathbf{0}$ and correspond to the valence band offset E_{vb}. The bands that do not contribute to the highest valence or the lowest conduction bands, $|\Psi_l\rangle$, are treated as second-order perturbation terms. The term $\frac{\hbar}{m_0}\mathbf{k}\langle\Psi_s|\hat{\mathbf{p}}|\Psi_s\rangle$ can be set to zero. This results from the fact that the s-like state has even parity and the operator $\hat{\mathbf{p}}$ has odd parity under inversion [144] and thus changes the parity of a wave function to which it is applied from odd to even and vice versa. Eq. (2.27) now reduces to:

$$\hat{H}_{11} = E_{cb} + \frac{\hbar^2 \mathbf{k}^2}{2m_0} + \frac{\hbar^2}{m_0^2} \sum_{i,j=x,y,z} k_i k_j \sum_{l \neq s, p_x, p_y, p_z} \langle \Psi_s | \hat{p}_i | \Psi_l \rangle \langle \Psi_l | \hat{p}_j | \Psi_s \rangle \frac{1}{E_{cb} - \varepsilon_l}, \quad (2.28)$$

where the $\hat{p}_i$'s are the momentum operators along $i = x, y, z$. Only few combinations of i, j and the bands Ψ_s, Ψ_{p_x}, Ψ_{p_y} and Ψ_{p_z} have a non-zero second order perturbation term due to symmetry reasons:

- While Ψ_s has odd parity in all directions, Ψ_j for $j = p_x, p_y, p_z$ is even in j-direction and odd in the other two directions.

- The operator $\hat{p}_j$ has odd parity along j and even parity along all other directions. $\hat{p}_j | \Psi_s \rangle$ therefore has even parity in j-direction and odd parity in all other directions.

- $\langle \Psi_s | \hat{p}_i$ has even parity along i and odd parity along all other directions. Thus, non-zero $\langle \Psi_s | \hat{p}_i | \Psi_l \rangle \langle \Psi_l | \hat{p}_j | \Psi_s \rangle$ terms only occur for $i = j$, since $\langle \Psi_l |$ and $| \Psi_l \rangle$ have the same symmetries.

In a cubic lattice, the x, y and z terms can be treated similarly, leading to a further simplification of Eq. (2.28) to:

$$\hat{H}_{11} = E_{cb} + \frac{\hbar^2 \mathbf{k}^2}{2m_0} + \frac{\hbar^2 \mathbf{k}^2}{m_0^2} \sum_{l \neq s, x, y, z} \frac{|\langle \Psi_s | \hat{p}_x | \Psi_l \rangle|^2}{E_{cb} - \varepsilon_l}, \quad (2.29)$$

since $|\langle \Psi_s | \hat{p}_x | \Psi_l \rangle|^2 = |\langle \Psi_s | \hat{p}_y | \Psi_l \rangle|^2 = |\langle \Psi_s | \hat{p}_z | \Psi_l \rangle|^2$. Using

$$D' = \frac{\hbar^2}{2m_0} + \frac{\hbar^2}{m_0^2} \sum_{l \neq s, x, y, z} \frac{|\langle \Psi_s | \hat{p}_x | \Psi_l \rangle|^2}{E_{cb} - \varepsilon_l}, \quad (2.30)$$

the Hamiltonian element $\hat{H}_{11}$ can be written as $\hat{H}_{11} = E_{cb} + D' \mathbf{k}^2$.

Similar considerations for the other elements of the 4 × 4-Hamiltonian in Eq. (2.25) working with the lowest conduction band and the three highest valence bands for a zincblende system leads to the following Hamiltonian:

$$\hat{H}^{4\times 4} = \begin{pmatrix} E_{cb} & 0 & 0 & 0 \\ 0 & E_{vb} & 0 & 0 \\ 0 & 0 & E_{vb} & 0 \\ 0 & 0 & 0 & E_{vb} \end{pmatrix} + \quad (2.31)$$

$$\begin{pmatrix} D'\mathbf{k}^2 & iP_0 k_x & iP_0 k_y & iP_0 k_z \\ -iP_0 k_x & -A'k_x^2 - B'(k_y^2 + k_z^2) & -C'k_x k_y & -C'k_x k_z \\ -iP_0 k_y & -C'k_y k_x & -A'k_y^2 - B'(k_x^2 + k_z^2) & -C'k_y k_z \\ -iP_0 k_z & -C'k_x k_z & -C'k_y k_z & -A'k_z^2 - B'(k_x^2 + k_y^2) \end{pmatrix},$$

where the other matrix elements are:

$$-A' = \frac{\hbar^2}{2m_0} + \frac{\hbar^2}{m_0^2} \sum_{l \neq s, p_x, p_y, p_z} \frac{|\langle \Psi_x | \hat{p}_x | \Psi_l \rangle|^2}{E_{\rm vb} - \varepsilon_l}$$

$$-B' = \frac{\hbar^2}{2m_0} + \frac{\hbar^2}{m_0^2} \sum_{l \neq s, p_x, p_y, p_z} \frac{|\langle \Psi_x | \hat{p}_z | \Psi_l \rangle|^2}{E_{\rm vb} - \varepsilon_l}$$

$$-C' = \frac{\hbar^2}{2m_0} + \frac{\hbar^2}{m_0^2} \sum_{l \neq s, p_x, p_y, p_z} \left(\frac{\langle \Psi_x | \hat{p}_x | \Psi_l \rangle \langle \Psi_l | \hat{p}_y | \Psi_y \rangle}{E_{\rm vb} - \varepsilon_l} + \frac{\langle \Psi_x | \hat{p}_y | \Psi_l \rangle \langle \Psi_l | \hat{p}_x | \Psi_y \rangle}{E_{\rm vb} - \varepsilon_l} \right)$$

$$P_0 = -i\frac{\hbar}{m_0}\langle \Psi_s | \hat{p}_x | \Psi_x \rangle = -i\frac{\hbar}{m_0}\langle \Psi_s | \hat{p}_y | \Psi_y \rangle = -i\frac{\hbar}{m_0}\langle \Psi_s | \hat{p}_z | \Psi_z \rangle \tag{2.32}$$

The elements D', A', B', C' and P_0 can now be related to the effective mass m_e and the Luttinger parameters γ_i similar to the corresponding elements in the EBOM Hamiltonian in Eq. (2.17). Furthermore, the comparison between $\hat{H}^{4\times 4}$ and Eq. (2.17) shows that both methods are identical at $\mathbf{k} = \mathbf{0}$. While the EBOM is based on a perturbation theory of the matrix elements in Eq. (2.16) and the parameters D', A', B', C' and ν are derived from atomic orbitals located at the corresponding lattice positions (see Eq. (2.18)), the atomistic nature vanishes within the $\mathbf{k}\cdot\mathbf{p}$ model and the effect of the lattice periodic potential $V(\mathbf{r}) = V(\mathbf{r} + \mathbf{R})$ is captured by effective masses and the Luttinger parameters [146]. The $\mathbf{k}\cdot\mathbf{p}$ model is based on a perturbation theory of the Hamiltonian (see Eq. (2.25)).

A higher number of explicitly treated bands leads to a better description of the band structure also far off the Γ-point but increases the complexity and thus the computational costs as well. It is even possible to obtain realistic band structures throughout the Brillouin zone by employing a full-zone $\mathbf{k}\cdot\mathbf{p}$ Hamiltonian, as was done with 15-band [41], 24-band [194] or 30-band [199] $\mathbf{k}\cdot\mathbf{p}$ Hamiltonians, allowing an excellent reproduction of the band structure throughout the Brillouin zone. Of course, such a higher sophisticated model requires a higher number of material parameters for the description of the additional bands. For the investigation of optoelectronic properties of devices, where in particular the region around $\mathbf{k} = \mathbf{0}$ is important, the explicit treatment of the lowest conduction band and the three highest valence bands and a perturbative treatment of all other bands in a $\mathbf{k}\cdot\mathbf{p}$-model has proven to deliver a sufficiently accurate description for most realistic direct semiconductor systems [151, 224].

Taking the spin-orbit coupling into account and, consequently, expanding the 4×4 model to an 8×8 model, Bahder [13] formulates the $\mathbf{k}\cdot\mathbf{p}$-Hamiltonian for zincblende structures in a basis of the complex eigenfunctions:

$$|\Phi\rangle = \begin{pmatrix} i|\Phi_s \downarrow\rangle \\ i|\Phi_s \uparrow\rangle \\ -\frac{i}{\sqrt{6}}\left(|\Phi_{p_x}\downarrow\rangle + i|\Phi_{p_y}\downarrow\rangle\right) + i\sqrt{\frac{2}{3}}|\Phi_{p_z}\uparrow\rangle \\ \frac{i}{\sqrt{2}}\left(|\Phi_{p_x}\uparrow\rangle + i|\Phi_{p_y}\uparrow\rangle\right) \\ \frac{-i}{\sqrt{2}}\left(|\Phi_{p_x}\downarrow\rangle - i|\Phi_{p_y}\downarrow\rangle\right) \\ \frac{i}{\sqrt{6}}\left(|\Phi_{p_x}\uparrow\rangle - i|\Phi_{p_y}\uparrow\rangle\right) + i\sqrt{\frac{2}{3}}|\Phi_{p_z}\downarrow\rangle \\ \frac{-1}{\sqrt{3}}\left(|\Phi_{p_x}\uparrow\rangle - i|\Phi_{p_y}\uparrow\rangle - |\Phi_{p_z}\downarrow\rangle\right) \\ \frac{1}{\sqrt{3}}\left(|\Phi_{p_x}\downarrow\rangle + i|\Phi_{p_y}\downarrow\rangle + |\Phi_{p_z}\uparrow\rangle\right) \end{pmatrix}. \quad (2.33)$$

The electronic contribution without taking strain and piezoelectricity into account then reads:

$$\hat{H}^{8\times 8}_{\text{unstrained}} = \begin{pmatrix} \hat{H}_c & \hat{H}_s \\ \hat{H}_s^\star & \hat{H}_v \end{pmatrix} = \quad (2.34)$$

$$\begin{pmatrix} A & 0 & V^\star & 0 & \sqrt{3}V & -\sqrt{2}U & -U & \sqrt{2}V^\star \\ 0 & A & -\sqrt{2}U & -\sqrt{3}V^\star & 0 & -V & \sqrt{2}V & U \\ V & -\sqrt{2}U & -P+Q & -S^\star & R & 0 & \sqrt{\frac{3}{2}}S & -\sqrt{2}Q \\ 0 & -\sqrt{3}V & -S & -P-Q & 0 & R & -\sqrt{2}R & \frac{1}{\sqrt{2}}S \\ \sqrt{3}V^\star & 0 & R^\star & 0 & -P-Q & S^\star & \frac{1}{\sqrt{2}}S^\star & \sqrt{2}R^\star \\ -\sqrt{2}U & -V^\star & 0 & R^\star & S & -P+Q & \sqrt{2}Q & \sqrt{\frac{3}{2}}S^\star \\ -U & \sqrt{2}V^\star & \sqrt{\frac{3}{2}}S^\star & -\sqrt{2}R^\star & \frac{1}{\sqrt{2}}S & \sqrt{2}Q & -P-\Delta_{\text{so}} & 0 \\ \sqrt{2}V & U & -\sqrt{2}Q & \frac{1}{\sqrt{2}}S^\star & \sqrt{2}R & \sqrt{\frac{3}{2}}S & 0 & -P-\Delta_{\text{so}} \end{pmatrix},$$

where $\hat{H}_c$ and $\hat{H}_v$ describe the conduction and valence band states. $\hat{H}_s$ denotes the superposition between valence band and conduction band states. The matrix elements consist of the operators:

$$A = E_c - \frac{\hbar^2}{2m_0}\gamma_c\left(\mathbf{k}_x^2 + \mathbf{k}_y^2 + \mathbf{k}_z^2\right),$$

$$P = -E_v - \gamma_1\frac{\hbar^2}{2m_0}\left(\mathbf{k}_x^2 + \mathbf{k}_y^2 + \mathbf{k}_z^2\right),$$

$$Q = -\gamma_2\frac{\hbar^2}{2m_0}\left(\mathbf{k}_x^2 + \mathbf{k}_y^2 - 2\mathbf{k}_z^2\right),$$

$$R = \sqrt{3}\frac{\hbar^2}{2m_0}\left[\gamma_2\left(\mathbf{k}_x^2 - \mathbf{k}_y^2\right) - 2i\gamma_3\mathbf{k}_x\mathbf{k}_y\right],$$

$$S = -\sqrt{3}\gamma_3\frac{\hbar^2}{2m_0}\mathbf{k}_z\left(\mathbf{k}_x - i\mathbf{k}_y\right),$$

$$U = \frac{-i}{\sqrt{3}}P_0\mathbf{k}_z,$$

$$V = \frac{-i}{\sqrt{6}}P_0\left(\mathbf{k}_x - i\mathbf{k}_y\right),$$

where the modified Luttinger parameters γ_i can be derived from the original Luttinger parameters γ_i^L as:

$$\gamma_c = \frac{m_0}{m_e} - \frac{E_p}{3}\left(\frac{2}{E_g} + \frac{1}{E_g + \Delta_{so}}\right),$$

$$\gamma_1 = \gamma_1^L - \frac{E_p}{3E_g + \Delta_{so}},$$

$$\gamma_2 = \gamma_2^L - \frac{1}{2}\frac{E_p}{3E_g + \Delta_{so}},$$

$$\gamma_3 = \gamma_3^L - \frac{1}{2}\frac{E_p}{3E_g + \Delta_{so}}.$$

The Kane matrix element E_p is:

$$E_p = 2m_0\frac{P_0^2}{\hbar^2}. \tag{2.35}$$

The Luttinger parameters are related to the effective masses of the three highest valence bands by [238]:

$$\frac{m_0}{m_{hh}^{[001]}} = \gamma_1 - 2\gamma_2, \qquad \frac{m_0}{m_{lh}^{[001]}} = \gamma_1 + 2\gamma_2,$$

$$\frac{m_0}{m_{hh}^{[111]}} = \gamma_1 - 2\gamma_3, \qquad \frac{m_0}{m_{lh}^{[111]}} = \gamma_1 + 2\gamma_3,$$

$$\frac{m_0}{m_{hh}^{[110]}} = \frac{1}{2}(2\gamma_1 - \gamma_2 - 3\gamma_3) \qquad \frac{m_0}{m_{lh}^{[110]}} = \frac{1}{2}(2\gamma_1 + \gamma_2 + 3\gamma_3)$$

$$\text{and } \frac{m_0}{m_{so}} = \gamma_1 - \frac{E_p\Delta_{so}}{3E_g(E_g + \Delta_{so})}. \tag{2.36}$$

By setting $E_\mathrm{p} = 0$, the Hamiltonian elements U and V become zero, the conduction and valence band states decouple and a 6+2-band model can be obtained. Furthermore, the original Luttinger parameters are not modified within this model. The 4-band model expressed by the Hamiltonian from Eq. (2.31) which neglects spin-orbital coupling can be reproduced by setting $\Delta_\mathrm{so} = 0$. For a 3+1-band model both the Kane parameter E_p and the spin-orbital splitting are set to zero. In this case, the complex multi-band physics reduces for the lowest conduction band to a simple particle-in-a-box problem, providing only the effective mass and the conduction band potential for the electron.

The above described continuum formalism reduces the realistic underlying zincblende symmetry, which is C_{2v}-type, to an artificial fcc (C_{4v}) symmetry. This can lead to artificially degenerate states in semiconductor nanostructures due to interface effects and is typically of growing influence with decreasing characteristic dimensions of the system. A detailed discussion of the origin and impact of this reduced symmetry is provided in Sec. 2.3.5.

The essence of the envelope function approach is the spatial dependence of the employed material parameters. The parameter set containing the unmodified Luttinger parameters γ_i^L, the effective mass m_e, the Γ-point energy terms E_cb, E_vb and E_p as well as the spin-orbital coupling Δ_so are spatially dependent in real space.

The parameter set itself can be calculated by fitting a $\mathbf{k}\cdot\mathbf{p}$ band structure around the Γ-point to experimentally determined band structures or to highly accurate ab initio band structures calculated e.g. using the G_0W_0-method [202]. This allows a multiscale approach, where parameters obtained from highly accurate DFT calculations on a small length scale are used in a continuum-formulated system containing 10^5 to 10^7 atoms, as is the typical case for many experimentally observed nanostructures. Moreover, atomistic models such as the EBOM or ETBM can employ parameters obtained from the same DFT calculations, allowing the consistent investigation for a wide range of nanostructure dimensions. The eight band $\mathbf{k}\cdot\mathbf{p}$-Hamiltonian for wurtzite systems can be derived similar to the zincblende Hamiltonian and is provided in the Appendix.

2.3.2 Deriving the k · p parameters from a given band structure

The $\mathbf{k}\cdot\mathbf{p}$ parameters required for the calculation of the electronic structure of a system can be derived from a given band structure, as observed in experiment or derived from theoretical calculations. For this purpose, the eigenvalues of the $\mathbf{k}\cdot\mathbf{p}$-Hamiltonian are fitted via a least-squares minimisation to this band structure for a given set of $\mathbf{k}$-values [202]. In the case of the eight-band $\mathbf{k}\cdot\mathbf{p}$ Hamiltonian for the zincblende structure and a direct band gap (i.e. conduction band minimum and valence band maximum are at the same $\mathbf{k}$), a fitting of the parameters m_e, γ_1, γ_2, γ_3 and P_0 is required. The band gap E_g and the spin-orbit splitting Δ_so can be taken directly from the band structure and require no fitting. With $\varepsilon_n(\mathbf{k})$ being the nth eigenvalue of the $\mathbf{k}\cdot\mathbf{p}$-Hamiltonian and $E_n^\mathrm{BS}(\mathbf{k})$ the energy of the corresponding band in the given input band structure, a minimisation of

$$\sum_{i=1}^{N_\mathbf{k}} \sum_{n=1}^{N_\mathrm{bands}} \left(\varepsilon_n(\mathbf{k}_i) - E_n^\mathrm{BS}(\mathbf{k}_i)\right)^2 \qquad (2.37)$$

is performed via the above fitting parameters. Summations are done over the number of involved bands and the selected **k** vectors. A corresponding mathematica script for this task is provided in Appendix C. For the III-nitrides which are the materials of interest in this work, we have employed **k** · **p**-parameters that have been recently derived using such a procedure [202, 239].

2.3.3 Influence of strain and polarisation

Semiconductor nanostructures are typically grown in molecular beam epitaxy (MBE) or metal-organic chemical vapour decomposition (MOCVD) processes. A crystalline material is grown epitaxially on another crystalline material. The primary observed growth modes are the Frank-van der Merwe growth [79], the Stranski-Krastanov growth [110, 172, 225] and the Vollmer-Weber growth [236]. Fig. 2.2 shows the three different modes for the epitaxial growth of a nanostructure in a matrix crystal material.

- The Frank-van der Merwe growth process is a 2d-layer-by-layer growth, i.e. a monolayer has to be completed before the growth of the next monolayer on top starts.

- The Stranski-Krastanov growth occurs typically after a certain layer thickness is exceeded. The reason for this is the Asaro-Tiller-Grinfeld instability [9, 93] which predicts an epitaxially grown, lattice mismatched surface to be unstable against perturbations above a critical size [220]. Above a certain thickness (in the III-nitrides, these are commonly two or three monolayers), new, complete monolayers are energetically unfavourable due to the resulting elastic energy stored in such systems and an island-growth on top of this so-called wetting layer occurs.

- In the Vollmer-Weber growth mode, the interactions between adatoms are stronger than the interactions between the adatoms and the surface leading to an island-growth of the adatom material on top of the crystal surface directly without a wetting-layer.

In all these growth modes, strain occurs when a lattice mismatch is present in the interface region between the structure and the surrounding matrix material. Within the wurtzite III-nitrides, this effect implies a strain induced polarisation together with a spontaneous polarisation and causes the appearance of polarisation potentials. Additionally, strained lattice constants in bulk crystals influence the band gap via deformation potentials [17, 89, 187]. Experimental observations show that external stress applied to a bulk cubic semiconductor can modify the band gap and induce a splitting of the highest valence bands [31, 97, 123]. Strain and polarisation modify the electronic structure and, therefore, influence the opto-electronic properties in semiconductor nanostructures. The different aspects of strain and polarisation shall be explained in more detail within this section.

Strain and polarisation potential V_p enter the **k**·**p** Hamiltonian as additional contributions [13]:

$$\hat{H}^{8\times 8} = \hat{H}^{8\times 8}_{\text{unstrained}} + \hat{H}_{\text{strain}} + V_p \tag{2.38}$$

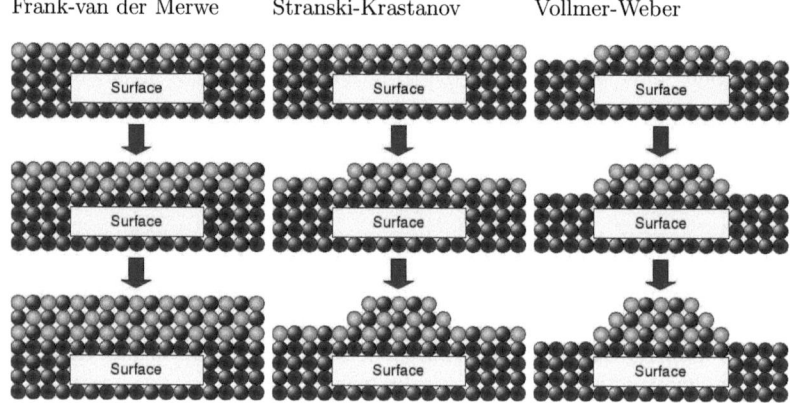

Figure 2.2: *Common modes in crystal growth. Brown and green spheres are the matrix crystal atoms, the yellow and brown spheres model nanostructure grown on top of the matrix material. Left: layer-by-layer Frank-van der Merwe growth. Middle: 3d island formation after 2d layer-by-layer growth: Stranski-Krastanov growth. Right: 3d island formation without previous 2d growth: Vollmer-Weber growth mode.*

$$\hat{H}_{\text{strain}} = \begin{pmatrix} a_c\epsilon & 0 & -v^\star & 0 & -\sqrt{3}v & \sqrt{2}u & u & -\sqrt{2}v^\star \\ 0 & a_c\epsilon & \sqrt{2}u & \sqrt{3}v^\star & 0 & v & -\sqrt{2}v & -u \\ -v & \sqrt{2}u & -p+q & -s^\star & r & 0 & \sqrt{3/2}s & -\sqrt{2}q \\ 0 & \sqrt{3}v & -s & -p-q & 0 & r & -\sqrt{2}r & 1/\sqrt{2}s \\ -\sqrt{3}v^\star & 0 & r^\star & 0 & -p-q & s^\star & 1/\sqrt{2}s^\star & \sqrt{2}r^\star \\ \sqrt{2}u & v^\star & 0 & r^\star & s & -p+q & \sqrt{2}q & \sqrt{3/2}s^\star \\ u & -\sqrt{2}v^\star & \sqrt{3/2}s^\star & -\sqrt{2}r^\star & 1/\sqrt{2}s & \sqrt{2}q & -a_v e & 0 \\ -\sqrt{2}v & -u & -\sqrt{2}q & 1/\sqrt{2}s^\star & \sqrt{2}r & \sqrt{3/2}s & 0 & a_v e \end{pmatrix} \quad (2.39)$$

with:

$$\begin{aligned} \epsilon &= \epsilon_{xx} + \epsilon_{yy} + \epsilon_{zz}, \\ p &= a_v(\epsilon_{xx} + \epsilon_{yy} + \epsilon_{zz}), \\ q &= b[\epsilon_{zz} - \frac{1}{2}(\epsilon_{xx} + \epsilon_{yy})], \\ r &= \frac{\sqrt{3}}{2}b(\epsilon_{xx} - \epsilon_{yy}) - i\tilde{d}\epsilon_{xy}, \\ s &= -\tilde{d}(\epsilon_{xz} - i\epsilon_{yz}), \end{aligned}$$

$$u = -i/\sqrt{3}P_0 \sum_j \epsilon_{zj}\mathbf{k}_j,$$
$$v = -i/\sqrt{6}P_0 \sum_j (\epsilon_{xj} - i\epsilon_{yj})\mathbf{k}_j,$$

where ϵ_{ij} is the spatially dependent strain tensor, b and $\tilde{d}$ are the shear deformation potentials and

$$a_v = \frac{dE_{\text{vb}}}{d\ln\Omega} \quad \text{and} \quad a_c = \frac{dE_{\text{cb}}}{d\ln\Omega} \tag{2.40}$$

with $d\ln\Omega = d\Omega/\Omega$ and Ω being the unit cell volume, are the hydrostatic valence band and conduction band deformation potentials [186, 234]. With these potentials, the change of the electronic band structure caused by the modified distances between the atoms in strained materials is taken into account. The matrix elements $\hat{H}^{ij}_{\text{strain}}$ can be seen as additional contributions of the electronic Hamiltonian elements $H^{8\times 8}_{\text{unstrained}}$ in Eq. (2.38) and are derived from solving the Schrödinger equation (Eq. (2.21)) for a strained coordinate system:

$$\left[\frac{\hat{\mathbf{p}}^2}{2m_0} + V(\mathbf{r}')\right]|\Psi_{n,\mathbf{k}}(\mathbf{r}')\rangle = \varepsilon_n(\mathbf{k})|\Psi_{n,\mathbf{k}}(\mathbf{r}')\rangle, \tag{2.41}$$

where the components of the strained coordinate system, $\mathbf{r}'$, are related to the unstrained coordinates by $r'_i = r_j \cdot (1+\epsilon_{ij})$ with $i,j = x,y,z$. From these considerations, a modification of the potential is derived such that [237]

$$V(\mathbf{r}') = V(\mathbf{r}) + \epsilon_{ij}V_{ij}(\mathbf{r}), \tag{2.42}$$

where the general $V_{ij}(\mathbf{r})$ potentials are related to the deformation potentials. The detailed considerations to determine the strain contributions to the Hamiltonian (Eq. (2.39)) are provided in Ref. [144].

2.3.4 Calculation of strain field and polarisation potential

The strain contributions that enter the $\mathbf{k}\cdot\mathbf{p}$-formalism via $\hat{H}_{\text{strain}}$ in Eq. (2.38) are calculated in a continuum model. In order to apply this model, one has to assume that the lattice mismatch of the involved materials is sufficiently small to allow a coherent interface. If this is not the case, misfit dislocations or even cracks will occur in a realistic system which is beyond the predictive capabilities of the continuum elasticity model considered here. While the strain fields calculated within this work are used as an input to calculate polarisation potentials and finally the electronic properties of semiconductor nanostructures, the presented formalism is not limited to semiconducting materials, but can be applied to various crystalline materials. The quantity to be minimised is the strain free energy:

$$F = \int d^3\mathbf{r} \frac{1}{2} C_{ijkl} [\epsilon_{ij} - \epsilon^{(0)}_{ij}][\epsilon_{kl} - \epsilon^{(0)}_{kl}], \tag{2.43}$$

$$\text{where} \quad \epsilon_{ij} = \epsilon_{ij}(\mathbf{r}) = \frac{1}{2}\left(\frac{\partial u_i}{\partial r_j} + \frac{\partial u_j}{\partial r_i}\right).$$

The strain tensor is calculated from the displacement fields u_i and the energy is minimised with respect to these displacement fields. $\epsilon_{ij}^{(0)}$ is a strain contribution resulting from the different bulk lattice constants of the involved materials:

$$\epsilon_{ij}^{(0)} = \begin{pmatrix} \frac{a_x - a_x^{(0)}}{a_x^{(0)}} & 0 & 0 \\ 0 & \frac{a_y - a_y^{(0)}}{a_y^{(0)}} & 0 \\ 0 & 0 & \frac{a_z - a_z^{(0)}}{a_z^{(0)}} \end{pmatrix}.$$

Where the a_i's are strain relaxed lattice constants and the $a_i^{(0)}$'s are the lattice constants of the unstrained materials at the according position. This results in $\epsilon_{xx,yy,zz}^{(0)} = (a - a^{(0)})/a^{(0)}$ for the cubic and $\epsilon_{xx,yy}^{(0)} = (a - a^{(0)})/a^{(0)}$ and $\epsilon_{zz}^{(0)} = (c - c^{(0)})/c^{(0)}$ for the hexagonal phase. The parameters $C_{ijkl} = C_{ijkl}(\mathbf{r})$ are the coefficients of the elastic tensor.

This leads to a set of three partial differential equations to be solved [190]:

$$\frac{\partial}{\partial x_i}\left(C_{ijkl}(\mathbf{r})\left[\frac{\partial u_k(\mathbf{r})}{\partial x_l} + \epsilon_{kl}^{(0)}(\mathbf{r})\right]\right) = 0 \text{ and } j = 1, 2, 3. \quad (2.44)$$

If no external pressure is applied to the system, it has to fulfill the following boundary condition at the cell boundaries:

$$n_i C_{ijkl}(\mathbf{r})\left(\frac{\partial u_k(\mathbf{r})}{\partial x_l} + \epsilon_{kl}^{(0)}(\mathbf{r})\right) = 0 \text{ for } j = 1, 2, 3, \quad (2.45)$$

where n_i is the normal vector of the corresponding boundary surface. This **second-order continuum elasticity model** can be modified by taking higher-order elasticity terms into account, e.g. in a **third-order** formulation [35, 230] where Eq. (2.43) is extended:

$$F = \int d^3\mathbf{r} \frac{1}{2!}C_{ijkl}[\epsilon_{ij} - \epsilon_{ij}^{(0)}][\epsilon_{kl} - \epsilon_{kl}^{(0)}] + \frac{1}{3!}C_{ijklmn}[\epsilon_{ij} - \epsilon_{ij}^{(0)}][\epsilon_{kl} - \epsilon_{kl}^{(0)}][\epsilon_{mn} - \epsilon_{mn}^{(0)}] \quad (2.46)$$

The third-order effects are supposed to produce only minor modifications of the electronic properties for realistic systems [183]. Indeed, in Sec. 4.1.2, the influence of third-order elasticity on the electronic properties of III-nitride quantum dots is shown to be negligible.

A displacement of atoms due to strain is the origin of **piezoelectric potentials**. In non-centrosymmetric crystals such as the wurtzite lattice, group theory predicts a non-vanishing bulk polarisation, i.e. the relaxed bulk crystal already shows **spontaneous polarisation** [122, 189, 197]. This is visualised in Fig. 2.3 for the example of a wurtzite structure. The center of gravity of the anion charges is not identical with the one of the cations. This introduces a dipole in the cell and causes the spontaneous polarisation. The piezoelectric polarisation $\mathbf{P}$ is commonly calculated from the strain fields:

$$\mathbf{P}_{\text{zincblende}} = \begin{pmatrix} e_{14}\epsilon_{xx} \\ e_{14}\epsilon_{yy} \\ e_{14}\epsilon_{zz} \end{pmatrix} \text{ and } \mathbf{P}_{\text{wurtzite}} = \begin{pmatrix} e_{15}\epsilon_{xz} \\ e_{15}\epsilon_{yz} \\ e_{31}(\epsilon_{xx} + \epsilon_{yy}) + e_{33}\epsilon_{zz} + \mathbf{P}_{\text{spont}} \end{pmatrix}, \quad (2.47)$$

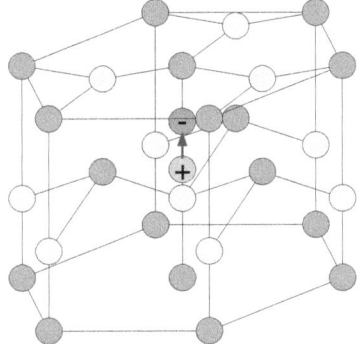

Figure 2.3: *A wurtzite lattice. Anions (blue) and cations (yellow) have different centers of gravity in the unit cell (marked with "-" and "+"). This causes the formation of a dipole (red) in the cell and leads to spontaneous polarisation even in a relaxed bulk crystal.*

where e_{14}, e_{15}, e_{31} and e_{33} are the piezoelectric constants that describe the response of mechanical stress to a present external electric field, or vice versa. $\mathbf{P}_{\text{spont}}$ is the spontaneous polarisation occuring in the wurtzite crystal structure [24]. The piezoelectric constants are known to be strain-dependent [14, 15, 16, 51, 235], i.e. these parameters in Eq. (2.47) are modified. For a wurtzite crystal, the piezoelectric constants e_{31} and e_{33} are given as [56]:

$$e_{31} = e_{31}^{(0)} + \frac{4eZ^*}{\sqrt{3}a_0^2}\frac{d\tilde{u}}{d\epsilon_{xx}} \quad \text{and} \quad e_{33} = e_{33}^{(0)} + \frac{4eZ^*}{\sqrt{3}a_0^2}\frac{d\tilde{u}}{d\epsilon_{zz}} \quad (2.48)$$

where Z^* is the Born effective charge, $\tilde{u}$ is the anion-cation bond length along the c-direction and a_0 is the bulk lattice constant. $e_{31}^{(0)}$ and $e_{33}^{(0)}$ are the unstrained piezoelectric constants. Within the present work, however, we stick to the common treatment of strain independent piezoelectric constants.

The piezoelectric potential $V_p(\mathbf{r})$ can be obtained from the piezoelectric polarisation $\mathbf{P}(\mathbf{r})$ via the Poisson equation:

$$4\pi\kappa_0 \nabla \left(\kappa_r(\mathbf{r})\nabla V_p(\mathbf{r})\right) = \varrho_p(\mathbf{r}). \quad (2.49)$$

Here, κ_0 and κ_r are the vacuum and the relative dielectric constants and $\varrho_p(\mathbf{r}) = -\nabla \mathbf{P}(\mathbf{r})$ is the polarisation charge density. Hence,

$$4\pi\kappa_0 \nabla \left(\kappa_r(\mathbf{r})\nabla V_p(\mathbf{r})\right) = -\nabla \mathbf{P}(\mathbf{r}) \quad (2.50)$$

$$\Longrightarrow \nabla V_p(\mathbf{r}) = -\frac{\mathbf{P}(\mathbf{r})}{4\pi\kappa_0\kappa_r(\mathbf{r})}. \quad (2.51)$$

The integration constant when integrating from Eq. (2.50) to Eq. (2.51) corresponds to a background polarisation and is already contained in the polarisation vector as $\mathbf{P}_{\text{spont}}$ in case it is non-zero. Similar to the material parameter set in the $\mathbf{k} \cdot \mathbf{p}$-formalism, the piezoelectric and elastic constants as well as the lattice constants are spatially dependent.

2.3.5 Artificial symmetries in the continuum picture

The investigation of geometrically ideal shaped quantum dots (e.g. square-based pyramids or lens-shaped quantum dots) in a continuum picture introduces artificial symmetries to the system, which are not observed with most atomistic methods. It is crucial to understand the origin and the impact of this effect in order to assess the reliability and accuracy of the applied continuum model. A truncated pyramidal GaN quantum dot in the zincblende structure is chosen to study this effect.

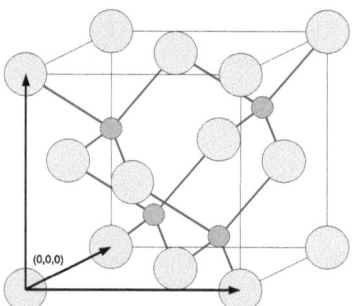

Figure 2.4: *The zincblende structure of InN, GaN and AlN. Green big circles represent In, Ga or Al atoms, blue small circles are the N atoms. Bonds are marked red. The rotation axes along (100), (010) and (001) are marked with black arrows.*

Even if strain and piezoelectric contributions are neglected, the atomistic setup around the interfaces gives rise to a reduction of the geometric C_{4v} (four-fold rotational) symmetry of the square-based truncated pyramid to a C_{2v} (two-fold rotational) symmetry, as shown by Bester and Zunger [29]. The C_{2v}-nature of the zincblende structure (Fig. 2.4) results from the possible rotations along the (100), (010) and the (001) axes in Fig. 2.4. Only twofold rotations (i.e. rotations of π) are rotations that reproduce the original crystal. Within the employed $\mathbf{k} \cdot \mathbf{p}$ model, similar effective masses and luttinger parameters along all three axes produce an artificial fourfold C_{4v} symmetry, which means that symmetry concerning rotations of $\pi/2$ is likewise assumed.

Figure 2.5 illustrates the origin of the reduced symmetry in a square-based, truncated pyramidal quantum dot in a zincblende lattice (see also Ref. [29]). For the top and base interfaces [001] and [00$\bar{1}$] it can be seen, that the anions inside and outside the quantum dot align in either the [1$\bar{1}$0] or the [110]-direction making these two directions energetically inequivalent. A larger area of the [001]-interface region is given in Fig. 2.6 to illustrate this alignment: The In atoms below a nitrogen always form chains along [1$\bar{1}$0], the nearest neighbouring Ga atoms above a nitrogen are always found in the [110]-direction. A similar behaviour can be seen for the side facets increasing this effect. In a continuum approach the [1$\bar{1}$0] and the [110] directions are treated identically, in an atomistic model one can see that the symmetry is actually lower.

In a continuum approach, this energetical equivalence of the [110] and the [1$\bar{1}$0] direction leads to artificially degenerate p- and d-like states. Even in the case of quantum wells, the

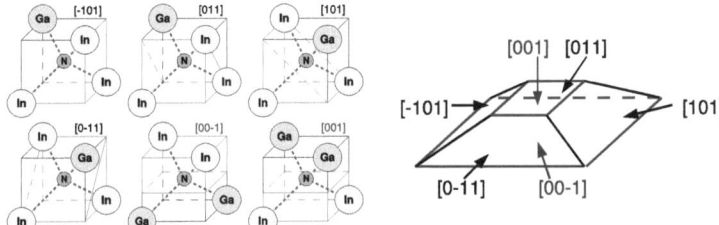

Figure 2.5: *Different interfaces between InGaN dot area (marked red) and GaN matrix (marked black). Plotted are the situation for the $[\bar{1}01]$, $[011]$, $[101]$, $[0\bar{1}1]$, $[00\bar{1}]$ and $[001]$ interfaces (left), green dashed lines represent the bonds between the atoms. The corresponding quantum dot is shown on the right side. Top and bottom interfaces are marked blue in the quantum dot sketch.*

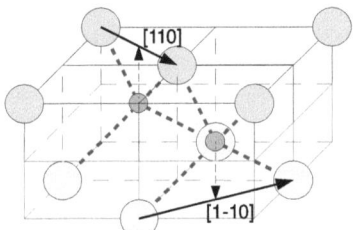

Figure 2.6: *A detailed view of the $[001]$ interface region. In atoms (yellow) below the nitrogens (blue) always form chains along the $[1\bar{1}0]$ direction, Ga atoms (green) above always line up along $[110]$. Bonds are marked in green dashed lines.*

two opposing interfaces are energetically different, as can be concluded from the $[00\bar{1}]$ and the $[001]$ interfaces in Fig. 2.5. Due to the absence of additional interfaces in an infinite quantum well, however, these effects compensate each other in such a system [57].

The effect of the atomistic nature of interfaces on the electronic properties is expected to decrease with growing system size, since the wave functions typically have less contact with the interface regions in larger systems. While the aspect discussed above holds even without taking atomic relaxation and strain effects into account, the effect is expected to increase in lattice-mismatched nanostructures when strain relaxation is present. Additionally, the fact that most realistic quantum dots are larger at the base than at the top leads to an increasing strain along the dot's growth axis. A single nitrogen anion therefore has two neighbouring indium cations below oriented along $[1\bar{1}0]$ direction and two neighbouring indium cations above oriented in $[110]$ direction. Due to the strain increasing along $[001]$-direction, the $[110]$-oriented cations feel a stronger strain than the $[1\bar{1}0]$-oriented ones. This increases the anisotropy of the $[110]$ and the $[1\bar{1}0]$ directions and leads to a further splitting of p- and d-like states [29]. A continuum elasticity model is in principle unable to consider this effect. In case these effects are expected to become non-negligible, strain fields obtained from atomistic calculations, e.g. from valence force field (VFF) methods [120], can be used as an input for the $\mathbf{k} \cdot \mathbf{p}$-model.

The displacement field resulting from atomistic relaxation leads to a piezoelectric potential in response [95]. The strain contributions computed from continuum elasticity theory do not result in an energetical preference towards the $[1\bar{1}0]$ or the $[110]$ direction, but the resulting piezoelectric potential does not exhibit the same C_{4v} symmetry as the dot geometry. This results in an energy splitting of p- and d-like states. The influence of atomistic character of the interfaces and piezoelectric potential for a given GaN quantum dot in a zincblende lattice will be discussed in chapter 4.1.1. An investigation of various InAs/GaAs nanostructures [29] reveals the maximum splitting of the energetically lowest two p-like electron states to be in the order of less than 10 meV, which is negligible for most applications.

2.4 The quantum confined Stark effect

A special focus of this work is on the influence of built-in electrostatic potentials on the electronic structure of semiconductor nanostructures, the quantum confined Stark effect.

The Stark effect is observed as a shift of wavelengths or a splitting of spectral lines of atoms or molecules when exposed to an external electric field [222]. It is therefore the electrostatic analogon to the Zeeman-splitting observed in the presence of an external magnetic field [257]. The Stark effect originates from the interaction of the electron charge density with the external field.

The **quantum confined Stark effect** is induced by applying an external electric field to a semiconductor structure with a given band gap embedded in a semiconducting matrix with another band gap [2, 160]. For nanostructures such as quantum dots, wires or wells, electrostatic potentials modify the energy levels and thus spectral lines in light emission processes. In the case of III-nitride quantum dots, e.g. GaN dots in AlN or InGaN dots in GaN, this leads to a reduction of the difference between electron and hole binding energies and thus to a redshift of the emission spectra. The presence of an electrostatic potential arising from an external field modifies the conduction and valence band edges in a semiconducting material. This modification leads to a localisation of electrons and holes on opposite sides of the structure. Fig. 2.7 sketches the nature of the quantum confined Stark effect. Semiconductor B has a small band gap and is embedded in material A which has a larger band gap. In the absence of external electric fields, electron and hole states can in principle localise at the same place. When an external electric field is introduced, the deformation of the conduction and valence band edge leads to a localisation of electrons on the left side of material B in Fig. 2.7 and of holes on the right side.

In non-centrosymmetric crystals, polarisation effects induce built-in electrostatic potentials. This effect is in particular strong in the III-nitride semiconductors and gives rise to the quantum confined Stark effect similar to an external electrostatic potential. This potential arises already from the spontaneous contribution P_{spont} in the polarisation in Eq. (2.47) in the wurtzite structure which is the common crystal structure for the III-nitrides. Additionally,

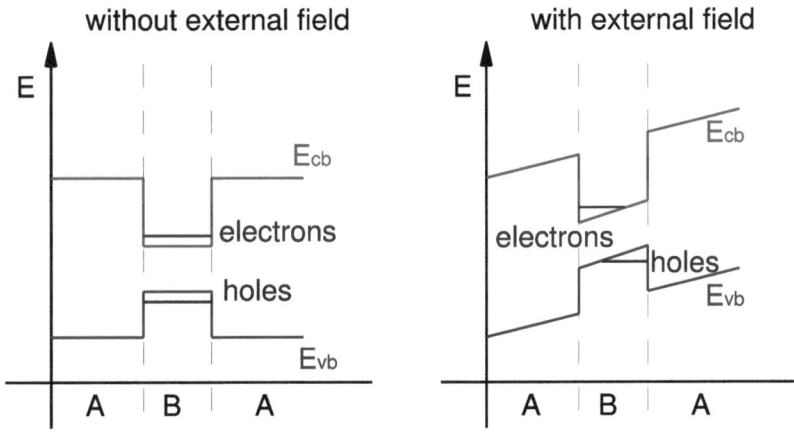

Figure 2.7: *Illustration of the quantum confined Stark effect. Left: semiconductor B embedded in A, no external fields. Right: Same situation when applying an external electric field. It can be seen, that electrons and holes localise at the opposite sides of material B.*

strain effects may increase the polarisation and thus the built-in electrostatic potentials. Within the zincblende phase, no spontaneous polarisation occurs and strain-induced contributions come up only in the presence of shear strains. Correspondingly, the quantum confined Stark effect is much larger in wurtzite systems, where it typically leads to a strong spatial separation of electrons and holes and therefore to poor excitonic recombination rates. Resulting piezoelectric potentials in wurtzite nanostructures are commonly larger than in zincblende structures of comparable size by a factor of ten [72]. The weaker potentials in typical zincblende nanostructures, e.g. in quantum dots, still do not justify to neglect the influence of shear-strain induced built-in potentials as they may lead to a reduction of the system's symmetry and thus to a splitting of artificially degenerated electronic states.

A detailed analytical example for the quantum confined Stark effect in a one-dimensional wurtzite system is provided together with the analysis of the electronic properties of In-GaN/GaN quantum wells grown on polar substrates in Sec. 4.3.1.

2.5 Emission and absorption spectra

The electron and hole wave functions and binding energies obtained from the above **k·p** model serve as an input for **many-particle calculations** to compute the absorption and emission spectrum of a semiconductor nanostructure. The calculation of these spectra is performed in the semiconductor theory group (Prof. F. Jahnke) at the Institute for Theoretical Physics, University Bremen and is thus not subject of this work. Nevertheless, the decisive steps to calculate the emission and absorption spectra of a nanostructure system shall be briefly explained for the sake of completeness.

2.5.1 Coulomb interaction

In a many-particle picture, the emission spectrum of a nanostructure is not determined purely by the electron and hole wave functions and binding energies. Electron and hole states influence each other via the **Coulomb interaction** of the involved charge [64, 65, 209]. In order to accurately determine the emission spectrum of a given system, it is therefore crucial to calculate the Coulomb matrix for the charge carrier states obtained from the $\mathbf{k} \cdot \mathbf{p}$ model.

The Coulomb interaction between the derived one-particle electron and hole states is defined as follows:

$$V_{ij,kl} = \int d^3\mathbf{r} \int d^3\mathbf{r}' \Psi_i^\dagger(\mathbf{r})\Psi_j^\dagger(\mathbf{r}')V(\mathbf{r}-\mathbf{r}')\Psi_k(\mathbf{r})\Psi_l(\mathbf{r}') \qquad (2.52)$$

$$\text{with} \quad V(\mathbf{r}-\mathbf{r}') = \frac{e^2}{4\pi\kappa_0\kappa_r|\mathbf{r}-\mathbf{r}'|} \quad \text{for} \quad \mathbf{r} \neq \mathbf{r}'$$

$$\text{and} \quad V(0) = \frac{1}{\Omega^2}\int_\Omega d^3\mathbf{r} d^3\mathbf{r}' \frac{e^2}{4\pi\kappa_0\kappa_r|\mathbf{r}-\mathbf{r}'|}.$$

Here, the $\Psi_{e/h}$ and $\Psi_{e/h}^\dagger$ denote the construction and destruction operators of the corresponding electron or hole wave function calculated using the $\mathbf{k} \cdot \mathbf{p}$-method, κ_0 and κ_r are vacuum and material dielectric constant, e is the electron charge and Ω is the unit cell's volume. The Coulomb matrix element $V_{ij,kl}$ quantifies the Coulomb interaction between a charge carrier (electron or hole) switching from state i to state l and a second carrier switching from state j to state k.

While the Coulomb matrix is required in a many-particle calculation of the emission and absorption spectrum of a given structure, it is not further employed within this work. For the investigation of charge carrier localisations, as performed for various structures in the present study, a simple **overlap matrix** is sufficient. Such an overlap matrix between electron and hole states can furthermore be used to provide an estimate of the recombination rate in semiconductor nanostructures [198]. This matrix quantifies the overlap of a ground state and excited states electrons and holes and is given as:

$$d_{ij} = \varrho_i^e(\mathbf{r})\varrho_j^h(\mathbf{r}) \quad \text{with} \quad \varrho_i^{e,h}(\mathbf{r}) = \langle \Psi_i^{e,h}(\mathbf{r})|\Psi_i^{e,h}(\mathbf{r})\rangle \qquad (2.53)$$

where the indices e and h denote the electrons and holes and i and j denote the eigenstate's numbers. For an eight-band wave function, a summation over the single components has to be done in order to compute the charge densities required for the calculation of the overlap matrix:

$$\varrho_i = \sum_{\sigma=0}^{8}\langle \Psi_\sigma^i|\Psi_\sigma^i\rangle. \qquad (2.54)$$

2.5.2 Many-particle Hamiltonian

When wave functions, binding energies and Coulomb matrix elements are known, a many-particle Hamiltonian in the envelope function approximation is used in order to calculate a nanostructure's emission spectrum [99]:

$$\hat{H} = \hat{H}_0 + \hat{H}_C, \qquad (2.55)$$

$$\hat{H}_0 = \sum_{i\sigma} \varepsilon_i^e e_{i\sigma}^\dagger e_{i\sigma} + \sum_{i\sigma} \varepsilon_i^h h_{i\sigma}^\dagger h_{i\sigma},$$

$$\hat{H}_C = \frac{1}{2} \sum_{\substack{ij,kl \\ \sigma,\sigma'}} V_{ij,kl}^{ee} e_{i\sigma}^\dagger e_{j\sigma'}^\dagger e_{k\sigma'} e_{l\sigma} + \frac{1}{2} \sum_{\substack{ij,kl \\ \sigma,\sigma'}} V_{ij,kl}^{hh} h_{i\sigma}^\dagger h_{j\sigma'}^\dagger h_{k\sigma'} h_{l\sigma} - \sum_{\substack{ij,kl \\ \sigma,\sigma'}} V_{ij,kl}^{eh} e_{i\sigma}^\dagger h_{j\sigma'}^\dagger h_{k\sigma'} e_{l\sigma}.$$

The indices i, j, k, l denote the charge carrier state, σ is the spin and e and h denote electron or hole wave functions. The Hamiltonian is split up in the non-interacting description of the single particle states calculated by the tight-binding or the $\mathbf{k} \cdot \mathbf{p}$-formalism $\hat{H}_0$ and the Coulomb interaction part $\hat{H}_C$. The resulting energy spectra can then be used to analyse and understand experimentally measured spectra from realistic nanostructured systems. The focus of this work, however, is on the investigation of single particle properties rather than on the optical properties resulting from a many-particle calculation. Eq. (2.55) is therefore given only for the sake of completeness.

Chapter 3

Plane-wave based implementation of the k · p-formalism and continuum elasticity theory

The relevant electronic states for optoelectronic applications based on semiconductor nanostructures such as quantum dots, wires and wells are typically strongly localised in real space. **k · p** and continuum elasticity models are, therefore, traditionally implemented in a finite elements scheme where gradient operators are realised using finite differences [223]. On the other hand, plane-wave approaches have proven to be a highly efficient tool to numerically solve differential equations such as, e.g., those considered in electronic structure codes employing density functional theory. One of the innovative aspects of the present work, which is explained in this chapter, is therefore, to use the plane-wave concept for an implementation of the continuum elasticity and the eight band **k · p** models discussed in the previous chapter [153].

A plane-wave based implementation has a number of advantages: First of all, **gradient operators** can be formulated much simpler in reciprocal space and require a significantly smaller computational effort than real-space finite difference operators do. The accuracy of the calculation can be directly controlled via the **number of plane-wave basis functions** taken into account. Furthermore, **available minimisation algorithms** of high efficiency can be directly employed with only a few modifications when incorporating these formalisms in an existing plane-wave code.

3.1 Plane-wave implementation of the eight band k · p formalism

The eight band **k · p** model from Sec. 2.3 has been implemented using a plane-wave formulation in the S/PHI/nX software library [219]. The plane-wave implementation of the zincblende Hamiltonian in Eq. (2.34) is demonstrated in this chapter, the wurtzite Hamiltonian in the Appendix (Eq. (A.1)) is implemented in a similar manner.

A function in real space that fulfills the periodic boundary condition $f(\mathbf{r}) = f(\mathbf{R} + \mathbf{r})$, where $\mathbf{R}$ is a linear combination of the vectors forming the periodic super cell, can be Fourier transformed to reciprocal space as:

$$\tilde{f}(\mathbf{G}) = \frac{1}{\sqrt{\Omega}} \int_\Omega f(\mathbf{r}) e^{i\mathbf{G}\mathbf{r}} d\mathbf{r}, \tag{3.1}$$

where $\mathbf{G}$ is the reciprocal lattice vector. For equidistant grids in $\mathbf{r}$ and $\mathbf{G}$ with the same dimension $N_x \times N_y \times N_z$, the integral over the volume Ω in Eq. (3.1) can be replaced by a sum and the corresponding summation can be performed using the efficient Fast Fourier Transformation (FFT). Its computational effort goes with $N \log N$, where $N = N_x \cdot N_y \cdot N_z$ is the total number of grid points and N_x, N_y and N_z denote the number of grid points along the corresponding directions.

Before discussing the complex eight band Hamiltonian, a simple effective mass model is used to illustrate the general idea. In this model, the Schrödinger equation $\hat{H}|\Psi\rangle = \varepsilon|\Psi\rangle$ is solved using an effective mass Hamiltonian, formulated in the atomic unit system, where $\hbar = m_0 = 4\pi\kappa_0 = 1$, in real space:

$$\hat{H}(\mathbf{r}) = \left(\frac{1}{2m_e(\mathbf{r})}\nabla^2 + V(\mathbf{r})\right). \tag{3.2}$$

Now the operators

$$\hat{\mathcal{G}} = \langle \mathbf{G}|\mathbf{r}\rangle \quad \text{and} \quad \hat{\mathcal{R}} = \langle \mathbf{r}|\mathbf{G}\rangle \tag{3.3}$$

are introduced. These operators perform the Fourier transformations from real to reciprocal ($\hat{\mathcal{G}}$) and from reciprocal to real space ($\hat{\mathcal{R}}$). The Hamiltonian $\hat{H}(\mathbf{r})$ can then be written in a formulation using real-space properties and a gradient formulation in reciprocal space as:

$$\hat{H}(\mathbf{r}) = \frac{1}{2m_e(\mathbf{r})}\hat{\mathcal{R}}\mathbf{G}^2 + V(\mathbf{r}). \tag{3.4}$$

Applying this Hamiltonian on a wave function $|\Psi(\mathbf{G})\rangle$ in reciprocal space then reads:

$$\hat{H}(\mathbf{G})|\Psi(\mathbf{G})\rangle = \hat{\mathcal{G}}\left[\frac{1}{2m_e(\mathbf{r})}\hat{\mathcal{R}}\left(\mathbf{G}^2|\Psi(\mathbf{G})\rangle\right) + V(\mathbf{r})\hat{\mathcal{R}}|\Psi(\mathbf{G})\rangle\right]. \tag{3.5}$$

The gradient operator ∇, which is computationally expensive in a finite differences formulation, can be easily expressed as a factor $\mathbf{G}^2$ in a plane-wave formulation.

The more complex eight band $\mathbf{k} \cdot \mathbf{p}$ Hamiltonian from Eq. (2.34) is implemented in a similar way. In the following, only one diagonal, one off-diagonal operator in the valence band part and one of the CB-VB coupling operators are described in detail. The application of the first diagonal operator of the Hamiltonian in Eq. (2.34) on the first component of an eight-component wave function $|\Psi\rangle$, can be considered in analogy to the effective mass situation discussed in Eq. (3.2):

$$\hat{H}_{11}|\Psi_1(\mathbf{G})\rangle = \hat{A}(\mathbf{G})|\Psi_1(\mathbf{G})\rangle = \hat{\mathcal{G}}\left[\frac{1}{2}\gamma_c(\mathbf{r})\hat{\mathcal{R}}\left(\mathbf{G}^2|\Psi_1(\mathbf{G})\rangle\right) + E_{\text{cb}}(\mathbf{r})\hat{\mathcal{R}}|\Psi_1(\mathbf{G})\rangle\right]. \tag{3.6}$$

All other diagonal elements are implemented correspondingly, $\Delta_{\text{so}}(\mathbf{r})$ behaves like the potential term $E_{\text{cb}}(\mathbf{r})$.

The Hamiltonian matrix element $\hat{H}_{35}|\Psi_5(\mathbf{G})\rangle = \hat{R}(\mathbf{G})|\Psi_5(\mathbf{G})\rangle$ is an off-diagonal term in the valence band part in Eq. (2.34):

$$\hat{H}_{35}|\Psi_5(\mathbf{G})\rangle = \hat{R}(\mathbf{G})|\Psi_5(\mathbf{G})\rangle$$

$$= \hat{\mathcal{G}}\left[\frac{\sqrt{3}}{2}\left[\gamma_2(\mathbf{r})\left(\hat{\mathcal{R}}(\mathbf{G}_x^2|\Psi_5(\mathbf{G})\rangle) + \hat{\mathcal{R}}(\mathbf{G}_y^2|\Psi_5(\mathbf{G})\rangle)\right) - 2i\gamma_3(\mathbf{r})\hat{\mathcal{R}}(\mathbf{G}_x\mathbf{G}_y|\Psi_5(\mathbf{G})\rangle)\right]\right], \quad (3.7)$$

all other off-diagonal elements of the valence-band related part of the Hamiltonian are formulated in a similar way. The coupling between conduction and valence band is introduced via the operators $U(\mathbf{r})$ and $V(\mathbf{r})$ in the eight band $\mathbf{k}\cdot\mathbf{p}$ Hamiltonian. As an example, the element $\hat{H}_{17}|\Psi_7(\mathbf{G})\rangle = U(\mathbf{G})|\Psi_7(\mathbf{G})\rangle$ is given by:

$$\hat{H}_{17}|\Psi_7(\mathbf{G})\rangle = U(\mathbf{G})|\Psi_7(\mathbf{G})\rangle = \hat{\mathcal{G}}\left[\frac{1}{\sqrt{3}}P_0(\mathbf{r})\hat{\mathcal{R}}(\mathbf{G}|\Psi_7(\mathbf{G})\rangle)\right], \quad (3.8)$$

and the other CB-VB-coupling elements are implemented correspondingly.

The indices i in the wave function $|\Psi_i\rangle$ denote the component of the full eight band wave function to be considered within a certain element of $\hat{H}|\Psi\rangle$. The multi-component wave functions $|\Psi\rangle$ are furthermore orthonormalised:

$$\langle\Psi^j|\Psi^{j'}\rangle = \sum_{\sigma=1}^{8}\langle\Psi_\sigma^j|\Psi_\sigma^{j'}\rangle\delta_{jj'}, \quad (3.9)$$

where j and j' denote the index of the corresponding electronic state and σ is the component of the eight band wave function.

3.2 Plane-wave based implementation of the continuum elasticity theory

The second order continuum elasticity model is implemented in an analogy to the $\mathbf{k}\cdot\mathbf{p}$ formalism. Applying an effective mass Hamiltonian or the Hamiltonian from Eq. (2.34) on a wave function $\hat{H}|\Psi\rangle$ in principle corresponds to determining the gradient $\delta F[u_x, u_y, u_z]/\delta u_i$ in terms of displacements u_i with $i = x, y, z$. Correspondingly, the gradient in Eq. (2.44) can be formulated in a plane-wave picture similar to $\hat{H}|\Psi\rangle$ in Eq. (3.5) as:

$$\frac{\partial F[u_x(\mathbf{G}), u_y(\mathbf{G}), u_z(\mathbf{G})]}{\partial u_i(\mathbf{G})} = \hat{\mathcal{G}}\sum_{j,k,l}\left[\hat{\mathcal{R}}\left(\hat{\mathcal{G}}\left\{C_{ijkl}(\mathbf{r})\left[\hat{\mathcal{R}}\{i\mathbf{G}_l u_k(\mathbf{G})\} + \epsilon_{kl}^0(\mathbf{r})\right]\right\}i\mathbf{G}_i\right)\right]. \quad (3.10)$$

For the electronic minimisation schemes (see next section), the similarity between $\hat{H}|\Psi\rangle$ and $\delta F[u_x, u_y, u_z]/\delta u_i$ for $i = x, y, z$ holds formally. Nevertheless, one has to be aware of two basic differences between wave functions and displacements:

- Eq. (2.44) is an inhomogeneous differential equation. This means that the displacements u_i must not be normalised, whereas the normalisation of the electronic wave function $\langle \Psi | \Psi \rangle = 1$ reflects the charge conservation.
- Electronic wave functions contain imaginary contributions and are no physical observables. The displacements u_i describe the difference between the strained position of a volume element and its original, unstrained position. Therefore, the u_i's do not contain any imaginary contributions.

It is important to note that a plane-wave implementation implicitly assumes **periodic boundary conditions**. For typical applications such as the investigation of single quantum dot systems, the cell size around the nanostructure has to be sufficiently large in order to prevent artificial interactions with periodically ordered neighbouring systems. Therefore, the cell size is a convergence parameter which has to be carefully checked. On the other hand, there is a number of experimentally observed systems where periodicity is explicitly given, such as superlattices and quantum dot arrays.

3.3 Electronic minimisation: The Conjugate-Gradient minimisation scheme

In order to find the correct solution of the Schrödinger equation

$$\hat{H} | \Psi(\mathbf{G}) \rangle = \varepsilon | \Psi(\mathbf{G}) \rangle \qquad (3.11)$$

or the elastic energy in Eq. (2.43) numerically, powerful minimisation techniques are required. These minimisation schemes are formally equivalent to the iterative minimisation algorithms applied in modern electronic structure codes to diagonalise the Hamiltonian. Therefore, the existing numerical tools can be easily reused. Within this work, the Conjugate-Gradient (CG) minimisation scheme [179] available within the S/PHI/nX program package has been adapted for both, the $\mathbf{k} \cdot \mathbf{p}$ formalism as well as for the continuum elasticity theory. The general algorithm as well as the necessary modifications for the two methods will be discussed next.

3.3.1 The basic Conjugate-Gradient algorithm

A single iteration step of the CG minimisation scheme to find the minimum energy ε of the Schrödinger equation (3.11) for the single-band effective mass Hamiltonian in Eq. (3.2) is shown in Tab. 3.1. The iterative minimisation is aborted by a set of termination conditions:

1. The difference between the ϵ values of two consecutive steps falls below a given convergence energy.
2. The value of θ defined in step 6 decreases below the numerical accuracy.

step		explanation
1	$\varepsilon^{(i)} = \langle \Psi^{(i)} | \hat{H} | \Psi^{(i)} \rangle$	Determine energy
2	$|\xi^{(i)}\rangle = \hat{H}|\Psi^{(i)}\rangle - \varepsilon^{(i)}|\Psi^{(i)}\rangle$	Calculate gradient
3	$|K\rangle = P(|\Psi^{(i)}\rangle)|\xi^{(i)}\rangle$	Apply preconditioning to gradient
4	$\text{tr}^{(i)} = \langle \xi^{(i)} | K \rangle$ $\gamma = \frac{\text{tr}^{(i)} \langle \xi^{(i-1)} | K \rangle}{\text{tr}^{(i-1)}}$ $|X^{(i)}\rangle = |K\rangle + \gamma |X^{(i-1)}\rangle$	Calculate search direction $|X^{(i)}\rangle$ from preconditioned gradient $|K\rangle$ and search direction of the previous iteration step $|X^{(i-1)}\rangle$.
5	$M = \begin{pmatrix} \varepsilon^{(i)} & \langle X^{(i)}|\hat{H}|\Psi^{(i)}\rangle \\ \langle \Psi^{(i)}|\hat{H}|X^{(i)}\rangle & \langle X^{(i)}|\hat{H}|X^{(i)}\rangle \end{pmatrix}$	Calculate Matrix M required for θ
6	$\theta = \frac{1}{2}\arctan\left(\frac{M_{10}+M_{01}}{M_{00}-M_{11}}\right)$	θ determines the contributions from previous wave function $|\Psi^{(i)}\rangle$ and search direction $|X^{(i)}\rangle$.
7	if $(M_{00} > M_{11})$ then $\theta = \theta - \pi/2$	Prevents finding a maximum instead of a minimum
8	$|\Psi^{(i+1)}\rangle = \cos\theta |\Psi^{(i)}\rangle + \sin\theta |X^{(i)}\rangle$	Compute updated wave function $|\Psi^{(i+1)}\rangle$

Table 3.1: *Conjugate gradient schemes for a single band effective mass model, which is adapted for an eight band* **k · p** *model and the continuum elasticity theory.*

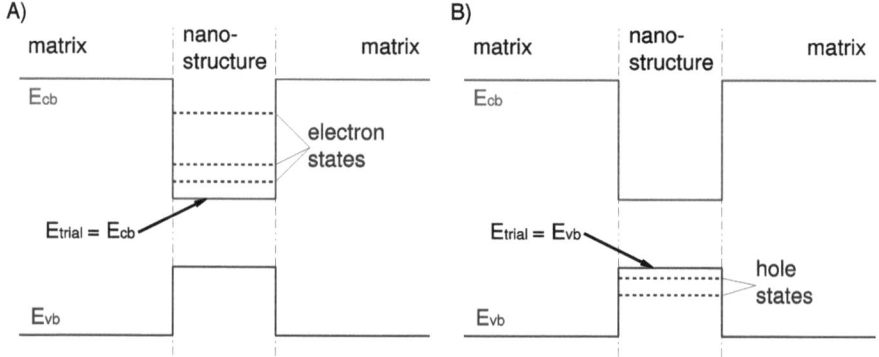

Figure 3.1: *For electron states, an energetical minimum above the conduction band has to be found (A) whereas hole states require a maximum below the valence band (B).*

Additionally, the minimisation is aborted without convergence if a given maximum number of convergence steps has been exceeded.

3.3.2 The CG implementation for the eight band k · p formalism

The basic minimisation scheme has to be modified to fit the requirements of the eight band **k · p** model. In particular, the basic algorithm is not designed for finding a maximum energy for the valence bands and a minimum for the conduction band simultaneously in case that all these bands contribute to a particular wave function (see Fig. 3.1). For the electron ground state, e.g., it is necessary to find a minimum energy above the conduction band minimum whereas the hole ground state has a maximum energy below the valence band maximum. A common solution for this problem is to minimise the squared difference between a given trial energy ε_t and the eigenenergy ε [223]. Therefore, the Hamilton operator and the energy must be substituted:

$$\hat{H} \longrightarrow (\hat{H} - \varepsilon_t)^2 \quad \text{and} \quad \varepsilon \longrightarrow (\varepsilon - \varepsilon_t)^2. \tag{3.12}$$

The gradient in step 2 in Tab. 3.1, therefore, is replaced by:

$$|\xi^{(i)}\rangle = \hat{H}^2|\Psi^{(i)}\rangle\langle\Psi^{(i)}|\hat{H}^2|\Psi^{(i)}\rangle - 2\epsilon_t^{(i)}\left(\hat{H}|\Psi^{(i)}\rangle\langle\Psi^{(i)}|\hat{H}|\Psi^{(i)}\rangle\right). \tag{3.13}$$

Additionally, matrix M (step 5) becomes:

$$M = \begin{pmatrix} (\varepsilon - \varepsilon_t)^2 & \langle(\hat{H} - \varepsilon_t)X|(\hat{H} - \varepsilon_t)\Psi\rangle \\ \langle(\hat{H} - \varepsilon_t)\Psi|(\hat{H} - \varepsilon_t)X\rangle & \langle(\hat{H} - \varepsilon_t)X|(\hat{H} - \varepsilon_t)X\rangle \end{pmatrix}. \tag{3.14}$$

Here, ε_t is a trial energy, which is an input parameter and has to be chosen according to the energy range of interest. If, e.g., the valence band edge is chosen as a value for ε_t, and the band gap is large enough that the first electron state has a much larger energy difference to ε_t than the hole states, the hole ground state is typically found first within the minimisation. Of course, this method requires an intelligent guess for the trial energy. Additionally, it happens under certain conditions that the choice of the conduction band edge for the value of ε_t still yields a determination of a hole state before electron states are found, if this hole state is energetically closer to the conduction band edge than the electron ground state. This occurs preferably in small band gap systems and can be avoided by choosing the trial energy sufficiently above the conduction band edge (see Fig. 3.1), however one has to be aware that choosing a too high trial energy might result in missing some of the lower lying electron states. Correspondingly, the trial energy is a parameter which has to be chosen carefully and in some cases needs to be adapted to the system of interest.

3.3.3 The CG implementation for the second-order continuum elasticity theory

In order to minimise the free energy in Eq. (2.43), the gradients $\delta F[u_x, u_y, u_z]/\delta u_i$ ($i = x, y, z$) are set to zero, as given in Eq. (2.44). Within our plane-wave implementation, the displacements u_i are treated similar to a wave function, while the gradient is used as a substitute for the Hamiltonian. Substituting $|\Psi\rangle$ by the displacements u_i and $\hat{H}$ by $\delta F[u_x, u_y, u_z]/\delta u_i$ and additionally setting $\varepsilon = 0$, one can directly apply the scheme described in Sec. 3.3.1. While $|\Psi\rangle$ in the eight band $\mathbf{k} \cdot \mathbf{p}$ model consists of eight components, u is treated similarly to a wave function with three components. The orthogonalization procedure applied to excited states in electronic minimisation becomes unnecessary for the strain field calculations.

3.3.4 Preconditioning

The eigenvalues of both the $\mathbf{k} \cdot \mathbf{p}$ Hamiltonian and the gradient in Eq. (2.44) consist of a $\mathbf{k}$-dependent term (kinetic energy and band interactions in the $\mathbf{k} \cdot \mathbf{p}$-formalism and all terms except those that contain $\epsilon_{ij}^0 \epsilon_{kl}^0$ in the continuum elasticity model) and a contribution which is independent of $\mathbf{k}$ (potential energy term and $\epsilon_{ij}^0 \epsilon_{kl}^0$ terms). Large $\mathbf{k}$-vectors introduced by a sufficiently large plane-wave cutoff and required for an accurate description of a complex given system lead to a poor energy convergence and therefore make preconditioning of the search direction X in step 4 of Tab. 3.1 necessary. Following Ref. [179], an appropriate plane-wave preconditioner for the conjugate gradient scheme in Sec. 3.3.1 is given by:

$$P = \frac{27 + 18\hat{x} + 12\hat{x}^2 + 8\hat{x}^3}{27 + 18\hat{x} + 12\hat{x}^2 + 8\hat{x}^3 + 16\hat{x}^4} \quad \text{with} \quad \hat{x} = \frac{\hat{D}}{\langle \Psi | \hat{D} | \Psi \rangle}. \tag{3.15}$$

An adaption to the $\mathbf{k} \cdot \mathbf{p}$ formalism, as suggested but not employed in Ref. [223], requires $\hat{D}$ to be the diagonal kinetic part of the eight band $\mathbf{k} \cdot \mathbf{p}$-Hamiltonian in Eq. (2.34). The effective mass is spatially constant, corresponding to the material where the wave function is expected to localise, which can be estimated from the conduction and valence band offsets. The similarity to a conventional plane-wave preconditioner where $\hat{D}$ is the Laplacian, ∇^2, is obvious. Please note that this preconditioner scheme requires the inverse operator of $\hat{D}$ in Eq. (3.15), which can be calculated straightforwardly within a plane-wave implementation. Tab. 3.2 shows the number of convergence steps required with and without preconditioning for an InN quantum dot embedded in a GaN matrix. The preconditioner decreases the number of required convergence steps by about a factor of 4.

For the continuum elasticity model, the same preconditioner can be employed. In this case, the operator $\hat{D}_i$ with $i = x, y, z$ becomes:

$$\hat{D}_i = -C_{iiii}\mathbf{G}^2. \tag{3.16}$$

For the continuum elasticity model, the preconditioner leads to a speed-up of approx. a factor of 2. The elastic constants have, similar to the effective masses and Luttinger parameters, no spatial dependence in the preconditioner. For the continuum elasticity model, these parameters are averaged over the whole cell, since no specific localisation of strain in the cell is expected. However, choosing one of the involved materials elastic constants instead of the averaged value in the preconditioner induces only a minor increase of the required number of convergence steps.

a) Total number of steps

	without preconditioner	with preconditioner
electrons	409	106
holes	1257	288

b) Average steps per order of magnitude of ε

	without preconditioner	with preconditioner
electrons	51	14
holes	157	34

Table 3.2: *Averaged required number of minimisation steps per state (a) and number of steps per order of magnitude (b) for energy convergence below 10^{-10} Hartree with and without preconditioner.*

3.3.5 Time reversal symmetry

The spin-orbit splitting Δ_{so} is often neglected in the calculation of electronic properties of semiconductor nanostructures. If this simplification is made, the energetically ordered eigenspectrum contains pairs of degenerated electron and hole states where only the spin-up and spin-down components are exchanged. Hence, only half of the eigenstates, e.g., the odd states, need to be determined, since the other ones are obtained from reversing the spin. Even for non-zero splitting, the choice of a spin-reversed odd eigenstate $|\Psi_{n-1}\rangle$ as an initial guess reduces the number of required minimisation steps for the following even eigenstate $|\Psi_n\rangle$

significantly in comparison to a randomised initial guess. For the above InN dot in GaN, the average odd electron state requires 106 steps for an energy convergence below 10^{-10} Hartree. The following even electron state requires only 22 steps when using the previous state after a spin-reversion as initial guess. For the holes, the odd states take 288 steps in average, whereas the following even state requires only 80 steps.

3.4 Piezoelectric potential

Within a real-space formulation, the piezoelectric potential is calculated by solving the Poisson equation (2.49). This task is another example for the high efficiency that can be achieved within a plane-wave formulation. Eq. (2.51) is, in atomic units with $4\pi\kappa_0 = 1$, given by:

$$\nabla V_\mathrm{P}(\mathbf{r}) = -\frac{\mathbf{P}(\mathbf{r})}{\kappa_r(\mathbf{r})} = -\mathbf{P}_\kappa(\mathbf{r}). \tag{3.17}$$

This real-space formulation requires a Poisson solver in order to determine the piezoelectric potential $V_\mathrm{P}(\mathbf{r})$. A corresponding formulation in reciprocal space can be solved straightforwardly:

$$-\mathrm{i}\mathbf{G}V_\mathrm{P}(\mathbf{G}) = -\mathbf{P}_\kappa(\mathbf{G}) \implies V_\mathrm{P}(\mathbf{G}) = -\frac{\mathrm{i}\mathbf{G}\mathbf{P}_\kappa(\mathbf{G})}{\mathbf{G}^2}. \tag{3.18}$$

The derived potential $V_\mathrm{P}(\mathbf{G})$ can easily be Fourier-transformed to a real-space $V_\mathrm{P}(\mathbf{r})$ afterwards.

Chapter 4

Application of the eight-band k · p formalism to a large scale of nanostructured systems

It is nowadays possible to grow zero-, one-, or two-dimensional III-nitride nanostructures with various electronic properties, as required for a specific application. While this allows to design structures and characterise them afterwards in an experiment, theoretical investigations allow to predict the properties of systems which can be designed experimentally, but also of those systems that are currently not in the range of experimental capabilities.

This work focuses on a broad scale of III-nitride nanostructures that exhibit promising properties for future technical, in particular light emitting, applications. For this purpose, the efficient plane-wave based implementation of the **k · p** model together with the second-order continuum elasticity theory, is an ideal tool to cover such a wide range of systems and possible modifications.

After a detailed evaluation of the previously introduced continuum models, zero-, one-, and two-dimensional nanostructures are investigated with respect to their elastic and electronic properties within this chapter. The resulting effects of the electronic structure on the optical properties of the studied systems are discussed.

4.1 Quantum dots: Zero-dimensional charge carrier localisation

Semiconductor quantum dots allow a quantum confinement of charge carriers in all three dimensions. As a consequence, well defined energy bands occur in these structures. However, what makes them particularly promising for a wide range of applications is that the positions of these levels, and furthermore the overlap of the corresponding charge densities which is of major importance for the optical transitions, can be modified by tailoring the size and shape of the quantum dots. Beyond light emission [124, 107], such nanostructures are promising candidates for future application in quantum computing [66, 170], e.g. as single-electron transistors [18, 119], quantum logic gates [19, 30, 90] or as photon detectors [22, 128, 195]. Correspondingly, the optoelectronic properties of quantum dots have received much research interest within the past years.

Experimental investigations were carried out on a large scale of material systems and applications. In particular, InAs/GaAs and InGaAs/GaAs quantum dots have been subject of extensive studies with respect to growth and electro-optical properties [102, 233] as well as to the development of specific laser [137] or quantum computing applications [91]. Further experimental studies were performed on other semiconductor materials, e.g., on GaSb/GaAs [227] or GaP/InP based quantum dots [173].

Theoretical studies of InAs [99, 245], and InGaAs [208] quantum dots in GaAs, of GeSi quantum dots in Si [132], or of GaN/AlN [72, 211] as well as InGaN/GaN [251] quantum dots have been performed employing different approaches ranging from continuum effective mass [99] and $\mathbf{k} \cdot \mathbf{p}$ models [72, 208, 251] to more sophisticated atomistic ETBM [211] and EPM calculations [245, 262, 263].

III-nitride quantum dots are of special interest due to the electronic properties of the involved materials which in principle allow to span the whole emission spectrum from infrared to ultraviolet light. Many available experimental studies provide information about, e.g., growth [228, 248] or optical properties [182, 248].

The focus of this chapter is on the electronic properties of a selection of polar and nonpolar grown III-nitride quantum dots in the wurtzite and the zincblende crystal structure. In particular, electron and hole localisation and binding energies will be investigated and used to draw conclusions about the optical properties.

A detailed evaluation of the employed formalisms is performed to verify the validity of this approach. For this purpose, the electronic properties of an example quantum dot system obtained from the eight-band $\mathbf{k} \cdot \mathbf{p}$ model are compared to those calculated in atomistic approaches. The employed second-order continuum elasticity model is evaluated in a comparison to a third-order elasticity model. In the following investigation of different polar and nonpolar grown III-nitride quantum dots, special attention is paid to the influence of the strong built-in electrostatic potentials on the charge carrier localisation and binding energies.

4.1.1 Comparison of atomistic and continuum approach calculations

Continuum effective mass models [109, 155] and multi-band $\mathbf{k} \cdot \mathbf{p}$ approaches employing six [73, 191] or eight [111, 192] band approximations have been successfully used in the past to model electronic properties of semiconductor nanostructures. Calculated absorption spectra based on eight band $\mathbf{k} \cdot \mathbf{p}$ calculations, e.g., for InAs quantum dots have been found to be in excellent agreement with experimentally observed photoluminescence spectra [223]. Despite the success of these calculations in describing the optoelectronic properties of quantum dots, it is essential to evaluate the method for the specific physical quantities which are subject of this work. Of course, continuum models are unable to provide a complete description of effects arising from single atoms or vacancies or atomistic effects at material interfaces, as pointed out in Sec. 2.3.5. It is therefore of crucial importance to evaluate the chosen continuum models for well defined systems before proceeding with the investigation of quantum dot, wire and well systems.

In this chapter, a detailed comparison between different $\mathbf{k} \cdot \mathbf{p}$ models and atomistic ETBM and EBOM (see Sec. 2.2) calculations is performed for the example of a zincblende GaN quantum dot embedded in AlN [151]. It is known that many III-nitride quantum dots grow in the wurtzite phase, a structure which exhibits strong piezoelectric effects. For a comparison of atomistic and continuum models, however, the zincblende structure with its much smaller piezoelectric effects is more favourable. It allows to focus solely on the electronic properties derived from atomistic and continuum models without considering strain and polarisation effects. Furthermore, recent experiments demonstrate that GaN quantum dots can indeed exist in a cubic structure [62, 150]. The ETBM and EBOM calculations which we use here for the comparison have been obtained in a cooperation with Stefan Schulz and Daniel Mourad (University Bremen), respectively [151].

The model system

The model quantum dot is shown in Fig. 4.1. Quantum dots of comparable shape grown in a cubic structure have been observed by different groups [1, 62, 83, 88, 150]. Typical diameters of 13 nm and heights of 1.6 nm have been observed [1, 83]. Within this comparison, the dimensions of these structures have been artificially decreased in order to investigate the impact of continuum approximations on nanostructures consisting of only a few thousand atoms, i.e., the dimensions are in the order of twenty atoms or less per direction. This drives the continuum methods to their limit. If for these extreme conditions they still provide acceptable results in comparison to the atomistic models, we can safely assume that a continuum-like description provides an even better description for larger systems.

The nanostructures investigated within this work commonly consist of more than 10.000 atoms. The bottom base length of the model quantum dot has been chosen as b=7 nm, the height is h=1.75 nm. The wetting layer thickness is 0.22 nm, which corresponds to one monolayer of AlN. The top base length is 3.5 nm. The cell was discretised in a grid of $80 \times 80 \times 80$ mesh points. Convergence tests of binding energies with respect to the mesh

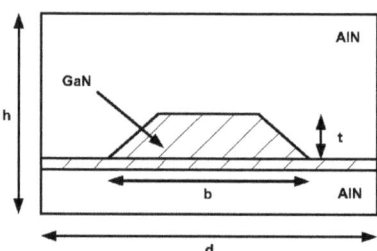

Figure 4.1: *Zincblende GaN dot in AlN matrix.*

accuracy can be found in the Appendix.

For wurtzite GaN/AlN systems, previous research revealed no intermixing of GaN and AlN within the wetting layer or the dot [248, 8]. Based on this observation, a similar behaviour in the considered zincblende system is assumed.

The $\mathbf{k} \cdot \mathbf{p}$ and tight-binding parameter set chosen for this comparison was obtained in Ref. [239] and is shown in Tab. 4.1. More recently calculated ab initio parameters applying the highly accurate $G_0 W_0$-method [202] do not produce important deviations to the GaN band structure around the Γ-point as obtained with the present set of parameters. Therefore, the comparison has not been repeated for these modified parameters.

	Parameter	GaN	AlN
lattice constant	a (Å)	4.5	4.38
band gap	E_{g} (eV)	3.26	4.9
valence band offset	ΔE_{vb} (eV)	0.8	0.0
X-point band energies	X_1^c (eV)	4.428	5.346
	X_3^v (eV)	-6.294	-5.388
	X_5^v (eV)	-2.459	-2.315
Kane parameter	E_{p} (eV)	25.0	27.1
Spin-orbit coupling	Δ_{so} (eV)	0.017	0.019
effective electron mass	m_e (m_0)	0.15	0.25
Luttinger parameters	γ_1	2.67	1.92
	γ_2	0.75	0.47
	γ_3	1.10	0.85

Table 4.1: *Material parameters for zincblende GaN and AlN [239, 80].*

Applied formalisms

The electronic structure of the chosen model system is calculated applying both atomistic and continuum theoretical models. The tight-binding method and the effective bond-orbital model are atomistic approaches while different $\mathbf{k}\cdot\mathbf{p}$ models ranging from effective mass approximations up to the eight band formalism derived in Chap. 2.3 use an envelope function approach.

For polar GaN and AlN, the valence band is formed by anions while the conduction band stems from the cations [185]. For the **empirical tight binding model** calculations, a $s_c p_a^3$-model is, therefore, assumed, the details of which are given in Ref. [210]. Within this approach the anions are modeled by their outer valence orbitals p_x, p_y and p_z, whereas the cations are described by an s orbital for each spin direction. The tight-binding parameter set is fitted to a recently calculated band structure [80] and can reproduce the Kohn-Luttinger parameters γ_i used within the $\mathbf{k}\cdot\mathbf{p}$ model.

Within the **effective bond-orbital model**, the tight-binding orbitals are replaced by the effective orbitals located on the sites of the underlying crystal lattice. Therefore, the original C_{2v} symmetry of the zincblende structure is raised to an artificial C_{4v} symmetry. The applied effective bond-orbital parametrization [143, 144] includes couplings up to the second-nearest neighbours in order to fit the bulk band structure to the considered set of $\mathbf{k}\cdot\mathbf{p}$ parameters. For further details, see Sec. 2.2.

The **eight band Hamiltonian** from Sec. 2.3.1 can be split up to lower levels of sophistication neglecting different physical effects. In this way an investigation of the importance of these effects becomes accessible. On the one hand, neglecting the Kane matrix parameter E_p yields a decoupled effective mass model for electrons and a six-band model for hole states. On the other hand, neglecting the spin-orbit coupling Δ_so reduces the eight band model to a four-band approximation and the effective mass and six-band model to a one-plus-three-band model.

Figure 4.2 shows the **bulk band structure for GaN** in the zincblende phase as computed by the continuum and the atomistic approaches. It can be clearly seen that all three major approaches show an excellent agreement around the Γ-point, due to the fitting of parameters in this region. Strong deviations for bigger $\mathbf{k}$-values are not relevant for the considered model system since the comparatively large dimensions of the nanostructure make the small $\mathbf{k}$-vectors more important. The spin-orbital coupling is not visible within this plot due to its small value of 0.017 eV in GaN.

Electron and hole states and binding energies

The absolute energies of electron and hole states obtained from atomistic methods and the eight band $\mathbf{k}\cdot\mathbf{p}$-formalism are shown in Tab. 4.2 and plotted in Fig. 4.3. The obtained values are in good agreement for the three methods. However, much more relevant for the optical properties are the energy differences between the excited and the ground states of electrons and holes. These differences are listed in Tab. 4.3. It is clearly visible that these energies

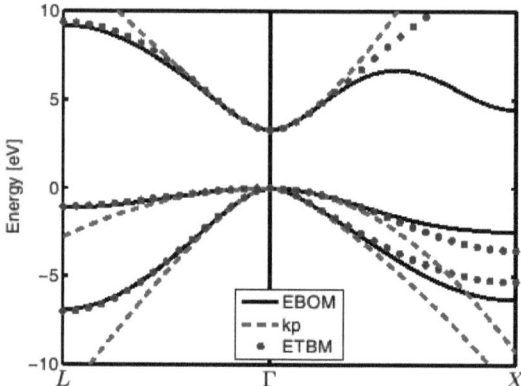

Figure 4.2: *Bulk band structure of zincblende GaN calculated using the tight-binding (dotted lines), effective bond-orbital model (solid) and the eight band* $\mathbf{k} \cdot \mathbf{p}$-*formalism (dashed lines).*

	$\mathbf{k} \cdot \mathbf{p}$	EBOM	ETBM
e_1 [eV]	4.4479	4.3429	4.3866
e_2 [eV]	4.5761	4.4623	4.5152
e_3 [eV]	4.5761	4.4623	4.5154
e_4 [eV]	4.5672	4.6284	4.6284
h_1 [eV]	0.6708	0.7244	0.6979
h_2 [eV]	0.6641	0.7189	0.6917
h_3 [eV]	0.6578	0.7145	0.6857
h_4 [eV]	0.6539	0.7123	0.6689

Table 4.2: *Single-particle energies for the truncated pyramidal GaN quantum dot.*

derived from the eight band $\mathbf{k} \cdot \mathbf{p}$-calculation agree with deviations of less than 1 meV with both atomistic models. The tight-binding results show a slightly lifted degeneracy of the 2nd and 3rd electron state, whereas these states are exactly degenerated within the effective bond-orbital model and the $\mathbf{k} \cdot \mathbf{p}$ calculation. This is due to the crystal symmetry which is described correctly as C_{2v} by the tight-binding approach only and is artificially raised to C_{4v} within the other two approaches (For details, see Sec. 2.3.5). However, this effect produces differences of only 0.2 meV between these states and will have only negligible influence on a resulting emission spectrum.

The charge densities of electrons (Fig. 4.4) and holes (Fig. 4.5) obtained from the eight band $\mathbf{k} \cdot \mathbf{p}$-calculation show an excellent qualitative agreement with those of the atomistic

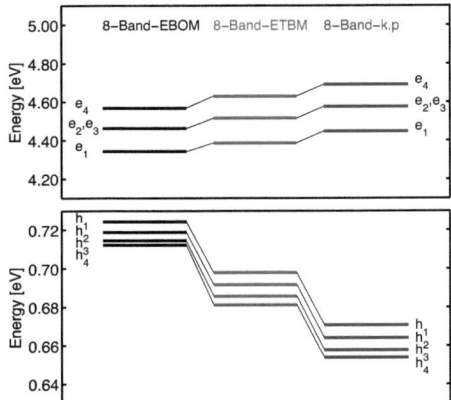

Figure 4.3: *Absolute energies of the first four electron (top) and hole (bottom) states obtained from the three approaches.*

	$\mathbf{k} \cdot \mathbf{p}$	EBOM	ETBM
$e_2 - e_1$ [eV]	0.1282	0.1194	0.1286
$e_3 - e_1$ [eV]	0.1282	0.1194	0.1288
$e_4 - e_1$ [eV]	0.2421	0.2855	0.2418
$h_2 - h_1$ [eV]	0.0067	0.0055	0.0062
$h_3 - h_1$ [eV]	0.0130	0.0099	0.0122
$h_4 - h_1$ [eV]	0.0169	0.0121	0.0167

Table 4.3: *Energies relative to electron and hole ground state.*

models. The above discussed artificial degeneracy of the 2nd and 3rd hole state can be found in these figures, too. While the empirical tight-binding, being able to resolve the correct crystal symmetry, finds a p_x^- and a p_y^- - like state, the other two approaches find states similar to $p_\pm = 1/\sqrt{2}(p_x + ip_y)$. Additionally, the crystal symmetry can be seen in the hole states h_3 and h_4 within the tight-binding results.

The effect of reduced $\mathbf{k} \cdot \mathbf{p}$-models can be seen in Tab. 4.4. Compared to the eight band model which can properly reproduce the atomistic results, simplified models employing a lower number of bands show large deviations in the quantitative or qualitative description of the electronic properties. Neglecting the spin-orbital coupling Δ_{so} within the 3+1 or 4-band approaches leads to an artificial degeneracy of the first two hole states which is not found in the atomistic models or the eight band results. A similar behaviour has been found in wurtzite InN/GaN quantum dots, recently [251]. It is important to note that this splitting of

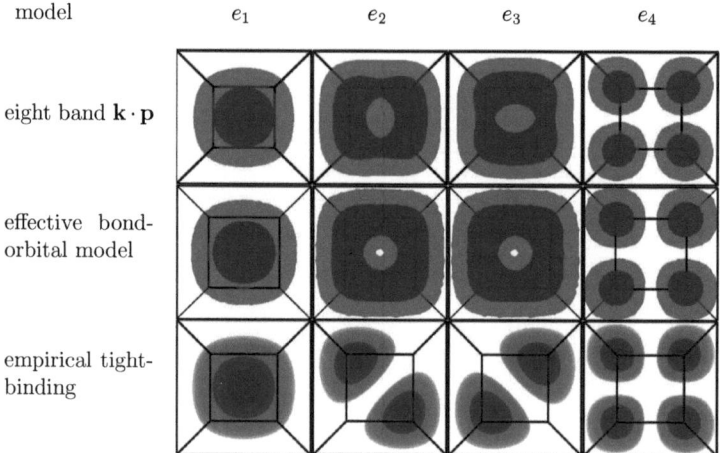

Figure 4.4: *First four electron states from* **k** · **p**, *effective bond-orbital model and empirical tight-binding. Depicted are isosurfaces of the probability density with 10% (red) and 50% (violet) of the maximum value.*

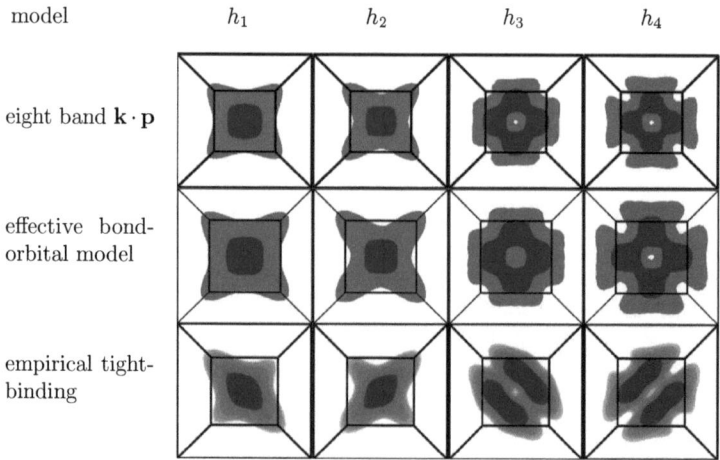

Figure 4.5: *First four hole states from* **k** · **p**, *effective bond-orbital model and empirical tight-binding.*

	3+1-band	4-band	6+2-band	8-band
e_1 [eV]	4.5259	4.4477	4.5259	4.4479
e_2 [eV]	4.6768	4.5759	4.6768	4.5761
e_3 [eV]	4.6768	4.5759	4.6768	4.5761
e_4 [eV]	4.8069	4.6897	4.8069	4.6900
h_1 [eV]	0.6679	0.6726	0.6677	0.6708
h_2 [eV]	0.6679	0.6726	0.6614	0.6641
h_3 [eV]	0.6622	0.6627	0.6570	0.6578
h_4 [eV]	0.6558	0.6591	0.6505	0.6539

Table 4.4: *Comparison of 3+1-band-, 6-band- and 8-band $\mathbf{k}\cdot\mathbf{p}$-results.*

the first two hole states is in the same order of magnitude as in comparable nanostructures that consist of semiconductor materials with a much higher spin-orbit splitting. For example, a splitting of 7 meV has been observed between the first two hole states in a CdSe pyramidal quantum dot, where the spin-orbit splitting has a value of 410 meV [210], whereas Δ_{so} has a value of only 17 meV in GaN and still induces a splitting of the first two hole states of 6.7 meV.

The $\mathbf{k}\cdot\mathbf{p}$ as well as the EBOM model allow to systematically investigate the influence of the spin-orbit splitting on the broken degeneracy of the first two hole states. In Fig. 4.6, the energy difference between the first two hole states is shown as a function of the spin-orbit splitting, which has been artificially rescaled from zero (which reflects the limit of a four-band model) to its correct value of 17 meV. The AlN spin-orbit splitting has been modified correspondingly, but has no significant influence on the electronic properties since the wave functions are localised within the GaN.

It can be seen that the energy difference between the two first hole states linearly depends on the spin-orbit splitting with a slope of 0.39. It can thus be concluded, that the small value of Δ_{so} in GaN does not justify to neglect this parameter in the calculation of electronic properties of GaN quantum dots. This result from the $\mathbf{k}\cdot\mathbf{p}$ calculation is also in excellent agreement with the corresponding EBOM calculation.

Decoupling the conduction and the valence band by setting the Kane matrix parameter $E_p = 0$ is often justified in the case of GaN by the large band gap of 3.26 eV [72]. Our investigations show that the hole states indeed remain almost unchanged by this modification. However, the relative electron state energies $e_2 - e_1$ and $e_3 - e_1$ show that this decoupling produces non-negligible deviations to the atomistic results as well as to the eight band model for electron states. This behaviour is observed despite the fact that GaN is a large-band gap material. The reason is the coupling between valence and conduction band due to the parameter P_0 and the related E_p (see Eq. (2.35)). This coupling leads to strong modifications of the Luttinger parameters and induces the coupling between $\hat{H}_c$ and $\hat{H}_v$. As a conclusion it is found that a decoupled 6+2-band model cannot be justified by the band gap alone, but rather by a small ratio of the Kane matrix parameter E_p and the band gap.

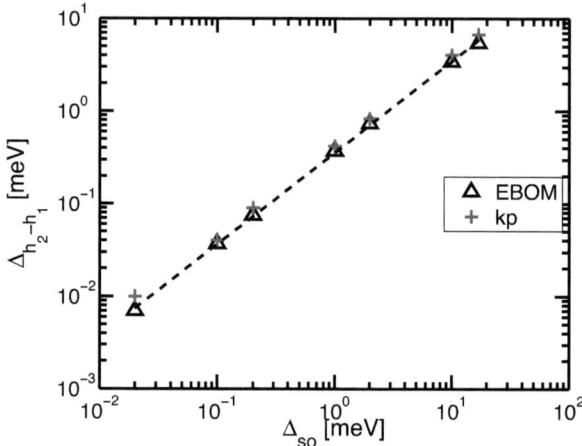

Figure 4.6: *Energy difference between the first two hole states as a function of the spin-orbit splitting Δ_{so} calculated using the EBOM (black triangles) and an eight-band $\mathbf{k}\cdot\mathbf{p}$ model (red crosses).*

Conclusions: Atomistic vs. continuum models

Summarizing the comparison of the atomistic ETBM and EBOM approaches with the continuum effective mass and $\mathbf{k}\cdot\mathbf{p}$ formalisms, a decoupled approach of the conduction band and the valence band contributions (six band Hamiltonian for holes and effective mass model for electrons) is found to induce a significant overestimation of the electron binding energies. Reducing the dimensions of the $\mathbf{k}\cdot\mathbf{p}$ Hamiltonian by neglecting the spin-orbit splitting in a four-band approach introduces an artificial degeneracy of the first two hole states in the present study. Furthermore, the energy difference between the first two hole states has been found to be a linear function of the spin-orbit splitting with a slope of 0.39 and thus non-negligible in comparable nanostructures, since the first two hole states are no longer energetically equivalent. It is furthermore most likely that similar splittings occur for energetically higher states which were not investigated within the present study.

If no such simplifications are made, the eight band $\mathbf{k}\cdot\mathbf{p}$ approximation has been found to be in excellent agreement with the atomistic models. The electron and hole binding energies show differences below 1 meV when comparing atomistic and continuum model calculations. The artificial degeneracy of the p-like electron states resulting from simplified symmetries in the continuum picture is found to be less than 0.2 meV in the atomistic ETBM picture and thus negligible for the purposes of this work.

The above comparison was done for a zincblende system. However, the agreement between atomistic and continuum descriptions in a wurtzite crystal lattice can be expected to be of similar quality, provided that the required parameter sets for the different methods are of similar consistency. In particular, artificial symmetries resulting from interface areas do not occur since the argumentation of Sec. 2.3.5 does not apply in this crystal structure. The large piezoelectric potential in a wurtzite nanostructure does not specifically break symmetries given by the underlying crystal structure and will therefore not induce deviations between atomistic and continuum descriptions.

The nanostructure studied in this section was assumed to have a diameter of 7 nm and a height of 1.75 nm, which is smaller than what is experimentally observed. The accuracy of a continuum model decreases when single atomistic effects become meaningful. Since the comparison at hand shows excellent agreement between atomistic and continuum models, larger structures of comparable complexity, where single atomistic effects become less important, can be described by a $\mathbf{k}\cdot\mathbf{p}$ model with even better agreement to atomistic calculations. Moreover, larger structures require higher computational effort for atomistic models whereas in a continuum model the computational expenses are similar to those of smaller structures.

In summary, the eight band $\mathbf{k}\cdot\mathbf{p}$ model is expected to provide a reliable and accurate description of the electronic properties of two and lower dimensional III-nitride nanostructures of comparable or larger characteristic dimensions than those studied here and will therefore be used for such studies within the following sections.

4.1.2 Influence of third-order elastic constants on electronic properties in III-nitride quantum dots

Within the previous section, strain effects were not taken into account, allowing a systematic comparison of the pure electronic atomistic and continuum descriptions of GaN quantum dots. For realistic systems, these contributions, of course, influence the electronic properties [213, 255, 259]. The continuum elasticity theory is commonly used to compute strain fields in semiconductor nanostructures [112, 113]. These strains enter the Hamiltonian matrix in Eqs. (2.38) and (A.1) and furthermore induce a polarisation potential via Eq. (2.47). Within this work, a second-order approximation of the elastic energy in Eq. (2.43) is made. In order to verify the validity of this approximation, the influence of higher-order terms has to be investigated. For this purpose, strain fields have been calculated employing only second-order elasticity theory and additionally using third-order terms expanding Eq. (2.43) to Eq. (2.46). The electronic properties are then compared with respect to the applied approximation of the continuum elasticity model.

The model quantum dot

Again, the material combination GaN/AlN was chosen. A wurtzite GaN quantum dot embedded in AlN was employed here to investigate the impact of third-order terms on the elastic and the resulting electronic properties. Within the wurtzite structure, the influence of polarisation potentials is typically stronger than in zincblende systems. Correspondingly, the effects of third-order terms in the minimisation of the elastic energy are expected to have more influence in wurtzite structures than in comparable zincblende systems. The model geometry is based on experimental high resolution transmission electron microscopy (HRTEM) observations [206]. The assumed shape of the quantum dot is a hexagonal, truncated pyramid with a height of 4 nm, a base diameter of 20 nm and a top diameter of 4 nm. The dot is grown on a wetting layer of 0.52 nm. The parameter set used for the calculation of strain, polarisation potential and the electronic properties is mainly taken from Ref. [72]. The second and third-order elastic constants were taken from Ref. [183].

Strain, polarisation and electronic properties

The strain components calculated using the second-order elasticity model are shown in Fig. 4.7 (left) along the [0001] center axis of the quantum dot. The strain fields for second and third order elasticity models have been calculated in cooperation with Toby D. Young (IPPT Warsaw) [67].

As an example for the differences between strain fields using second-order elasticity theory and those obtained from a third-order model, the diagonal strain component $\epsilon_{xx}^{(2)}$ from a second-order calculation and the difference $\epsilon_{xx}^{(3)} - \epsilon_{xx}^{(2)}$ is shown in Fig. 4.8. The differences are in the order of 10^{-6}, which is significantly smaller than the strain values in the order of 10^{-3}. For the other diagonal and off-diagonal strain components, the deviations are in the same order of magnitude and therefore not shown in additional plots. The resulting piezoelectric potential (Fig. 4.7 right), correspondingly, shows only minor modifications when taking third-order elasticity effects into account. These differences are shown in Fig. 4.9. It can be seen that these deviations are below 0.5 meV, i.e., below 0.1% of the absolute value.

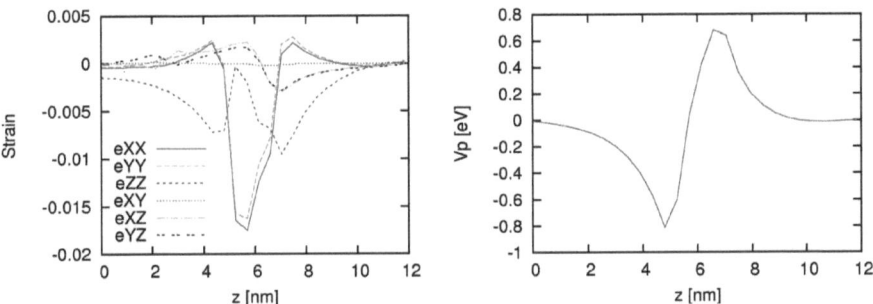

Figure 4.7: Left: Strain tensor elements ϵ_{xx}, ϵ_{yy}, ϵ_{zz}, ϵ_{xy}, ϵ_{xz} and ϵ_{yz} along the [0001] axis through the quantum dot center. Right: Polarisation potential along the same line.

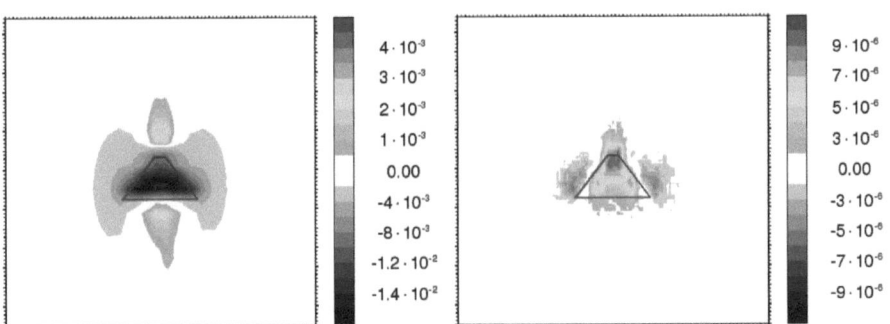

Figure 4.8: Left: Strain component ϵ_{xx} calculated using second-order elasticity theory, cut through the center of the dot in the x-z layer. The cell size is 80 nm along x and 12 nm along z direction. Right: Difference between third-order elasticity calculated strain ϵ_{xx} and the second-order calculation. In both plots, the quantum dot is depicted in red.

The first three electron and hole states are shown in Fig. 4.10. It can be seen that these results are in qualitative agreement with calculations performed in GaN/AlN QDs of similar shape (For example, see Ref. [72]). In comparison to a calculation without strain effects, already the strain fields calculated using the second-order model do not significantly alter the charge carrier localisation. Employing third-order elasticity effects does not introduce

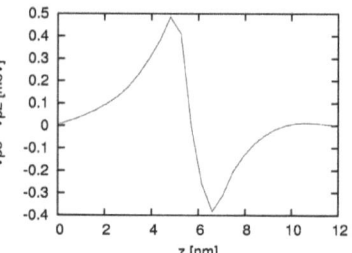

Figure 4.9: *Right: Difference between polarisation potential calculated with third-order elasticity theory and with only second-order effects in meV.*

any visible modifications in the charge carrier localisation. In particular, the character of the electronic states (e.g. the s-like character of the electron ground state or the p-like character of the two following electron states) is not modified. The binding energies of these electron and hole states are given in Tab. 4.5. Here, it can be seen that strain indeed leads to non-negligible modifications of the electronic structure. Employing third-order elasticity, however, leads to modifications in the order of 0.1 meV in the investigated system, which is a very small effect.

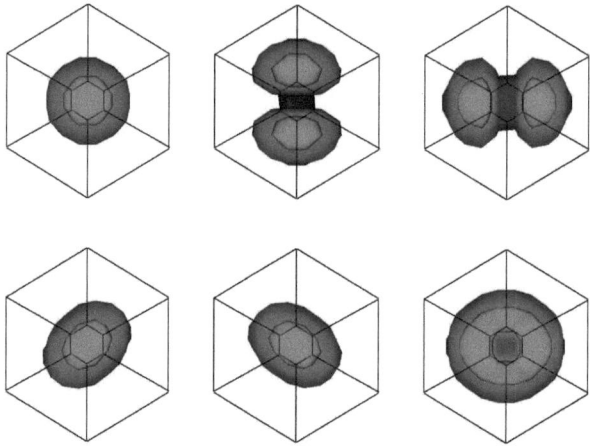

Figure 4.10: *Three lowest electron (top) and hole states (bottom) in a wurtzite GaN quantum dot. The corresponding binding energies are given in Tab. 4.5.*

	e_0	e_1	e_2	h_0	h_1	h_2
no strain	3.9779	4.1115	4.1161	0.0979	0.0979	0.0883
second-order	4.1921	4.3100	4.3201	0.0487	0.0480	0.0417
third-order	4.1923	4.3102	4.3203	0.0486	0.0480	0.0416

Table 4.5: *Binding energies of the first three electrons and holes in eV.*

Conclusions: Second and third-order elasticity

The influence of third-order elastic effects on strain, polarisation and electronic properties of a wurtzite GaN quantum dot in an AlN matrix has been studied. All investigated properties show only very small modifications when taking third-order elastic constants into account. The binding energies are modified by much less than 1 meV and, therefore, show that third-order elasticity contributions can be safely neglected for the studied quantum dot system. Third-order elasticity effects might become more important for larger strain values. Keeping in mind that for the description of epitaxial growth processes continuum elasticity is limited to small strain values (typically below 10% [144]), it can be concluded that third-order elasticity effects can be neglected in all systems studied within this work.

4.1.3 Influence of size, shape and material composition on optical properties of GaN quantum dots

The investigation of a wide range of modifications to a reference quantum dot system allows to design nanostructures for specific applications. Furthermore, such studies allow for a detailed understanding of the impact of the system's parameters. The $\mathbf{k}\cdot\mathbf{p}$ formalism is excellently suited to provide information about the influence of shape, size and material composition on the electronic properties of semiconductor nanostructures. Moreover, the continuum description of nanostructured systems which does not consider single atoms allows to use the converged electronic wave functions computed for one structure as an initial guess for various modifications of this system which dramatically reduces the computational effort in systematic studies of such modifications. The highly efficient plane-wave implementation introduced in Chap. 3 allows to achieve a high throughput of calculations, making studies of a wide range of possible modifications to a given quantum dot system possible.

In this section, different parameters of a GaN/AlN quantum dot are varied and allow a qualitative and quantitative understanding of the impact of these properties on the electronic structure. Based on experimental observations [248] and similar to the model system in the previous section, a hexagonal wurtzite GaN quantum dot with a base diameter of 20 nm and a height of 4 nm has been chosen as initial structure for this study. The calculations have been done considering polarisation and strain effects.

Quantum dot size

The initial quantum dot geometry was rescaled systematically from 0.25 of its original dimensions to 3 times the original height and diameter. Fig. 4.12 shows the dependence of the electron and hole eigenenergies on the quantum dot's size. A strong influence of the structure's characteristic dimensions on the electronic structure is visible. In the region below the size of the reference structure, quantisation effects induce a difference between ground and excited states. Moreover, it can be seen that the eigenenergies of the electrons decrease almost linearly above the dimensions of the reference structure. The hole states show a similar linear behaviour, but increasing in energy. This means that the electron and hole eigenenergies do not converge towards the conduction and valence band edges of the quantum dot material. This is a result of the built-in electrostatic potential which increases with the quantum dot's size (See Fig. 4.11). A detailed analysis of this effect will be given for polar InGaN quantum wells in Sec. 4.3.1.

The obtained results are in agreement with previous work on GaN/AlN quantum dots [7, 72]. A similar behaviour has also been found in InN/GaN quantum dots [251].

The number of confined electron states is shown as a function of the quantum dot size in Fig. 4.13. A small number of confined states is of particular importance for the application in single-photon emitters, whereas a high number of confined states increases the complexity of the absorption and emission spectra. It is nicely visible, that already for structures smaller than the initial model quantum dot a huge number of localised electrons is found. Due to its better localisation properties, the number of confined hole states is much higher. This

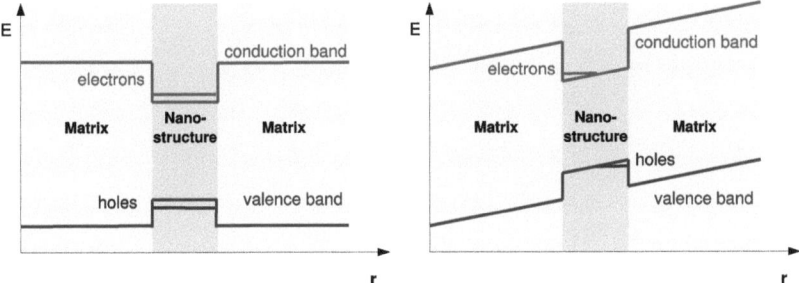

Figure 4.11: *Conduction and valence band offsets in nanostructures in absence (left) and presence (right) of a built-in electric field. If such an electric field is present, electrons (dark violet) and holes (light violet) localise at different positions inside the nanostructure.*

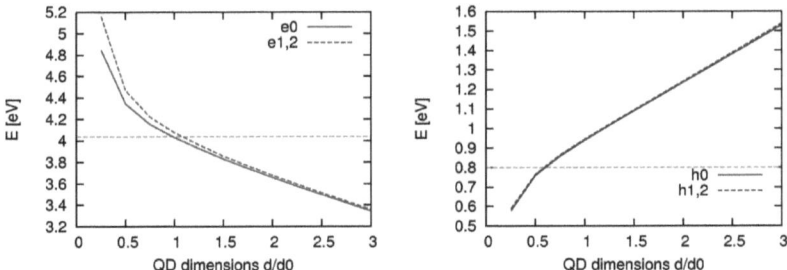

Figure 4.12: *Electron (left) and hole (right) eigenenergies in a GaN quantum dot as a function of the quantum dot's size. d_0 is the original dimension (height or diameter), d is the resized one. The green dashed line denotes the bulk GaN conduction band (left) and valence band (right) edge.*

number has not been explicitly calculated, due to the huge computational effort required for such a large number of hole states.

Quantum dot shape

The reference quantum dot is a truncated pyramid based on a regular hexagon. The influence of the quantum dot's shape was studied by assuming a regular basis of the truncated pyramid with a variable number of edges. Starting from a triangular basis as observed, e.g, in Ref. [81], the number of edges has been increased up to 20. The distance from the base center to the edges is kept constant, resulting in a slight volume increase of the quantum dot with the

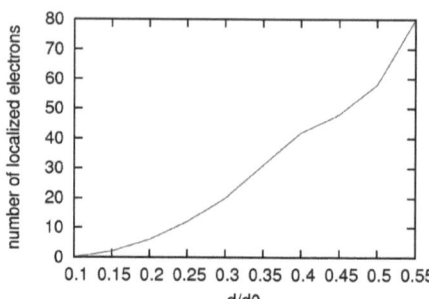

Figure 4.13: *Number of electron states localised inside the quantum dot as a function of the dot's characteristic dimensions.*

number of edges. The eigenenergies of the energetically lowest electron and hole states are shown in Fig. 4.14. It is nicely visible that the eigenenergies converge towards the limit of a circular basis. Moreover, the hexagonal based pyramid already exhibits electronic properties close to the limit of a circular based dot. The first and second excited electron states as well as the first and second excited hole states are energetically close to each other. The energy difference between the energetically close electron and hole states is shown in Fig. 4.15. It can be seen that, within the $\mathbf{k} \cdot \mathbf{p}$ model, the first two excited electron states are degenerate if the number of edges is a multiple of 4, which can be understood due to the p-like structure of these states. This degeneracy means, that these states are energetically equivalent. The situation is different for the hole states, where the energetical difference between the first and the second excited hole state decreases when the dot shape converges towards a circular geometry. Due to the spin-orbit splitting, however, no complete degeneracy of these states occurs.

Influence of the material composition

It is known from experiment, that pure GaN quantum dots in an AlN matrix show almost no atomistic interdiffusion around the interfaces [248]. Due to the small lattice mismatch between GaN and AlN, however, it is possible to modify the quantum dot band gap and thus to directly influence the electronic properties. In particular, the wavelength of emitted light can be controlled via the Ga-content in an $Al_{1-x}Ga_xN$ wurtzite quantum dot. The Ga content in an $Al_{1-x}Ga_xN$ quantum dot was varied from 0.1 to 1.0 in order to investigate the behaviour of the electron and hole state energies as a function of the material composition of the quantum dot. The AlN matrix remains unaffected. The common approach to determine the material parameters of ternary alloys is a linear interpolation between the bulk material parameters. Only for the band gap and the spontaneous polarisation a quadratic interpolation is done.

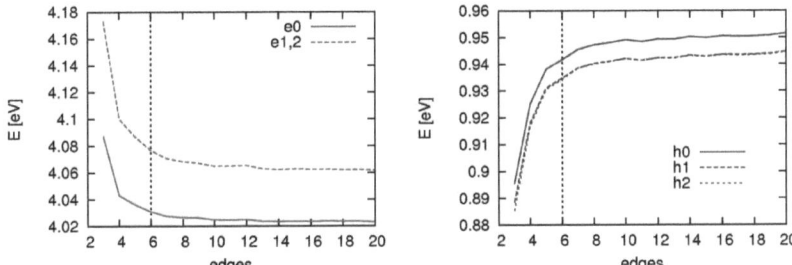

Figure 4.14: *Electron (left) and hole (right) eigenenergies in a GaN quantum dot for different basis polygons. The dashed vertical line denotes the experimentally observed hexagonal structure.*

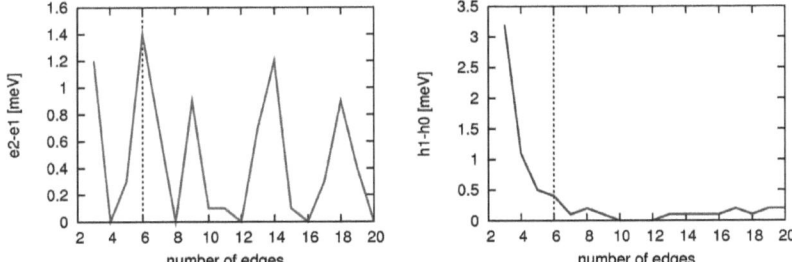

Figure 4.15: *Energy difference between the second and the first excited electron state (left) and the ground and first excited hole states (right) eigenenergies in a GaN quantum dot for different basis polygons. The dashed line indicates the hexagonal dot shape.*

Here, bowing parameters of $b_{E_g}(AlGaN) = 0.7$ eV and $b_{P_{sp}}(AlGaN) = -0.021$ C/m² have been chosen, as suggested in Ref. [239]. All other parameters can be found in Tab. 4.10 on page 100.

In Fig. 4.16 it can be seen that the electron and hole energies depend almost linearly on the Ga content x, indicating a weak influence of the bowing. This behaviour was predicted for InGaN/GaN quantum dots by Winkelnkemper et al. using an eight-band $\mathbf{k} \cdot \mathbf{p}$ model, recently [251], and allows, in principle, to derive and to reproduce the Ga content for an unknown AlGaN composition from the measured emission wavelength in technical applications or, vice versa, the choice of a composition to achieve a specific emission wavelength.

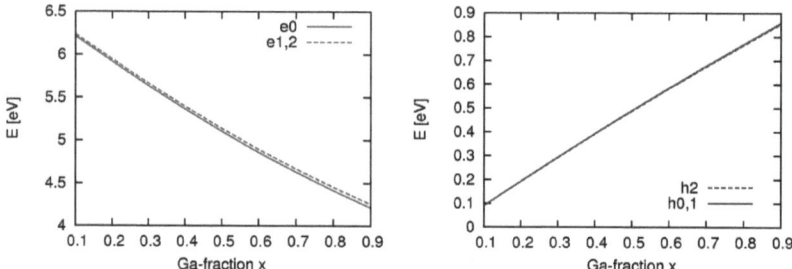

Figure 4.16: *Electron (left) and hole (right) state energies in an $Al_{1-x}Ga_xN$ quantum dot as a function of the Ga content x.*

Summary: Size, shape and material composition of III-nitride quantum dots

The electronic properties of a GaN/AlN quantum dot have been studied systematically for different sizes, shapes and material compositions. All of these properties were found to significantly modify the electronic structure of the quantum dots. The influence of the material composition, i.e., the GaN content in a ternary AlGaN quantum dot and the influence of the dot size were found to induce modifications in the eigenenergies in the order of eV, whereas the shape of the dot has a minor influence in the order of approx. 10 meV. It can be concluded that the optoelectronic properties of such nanostructures, in particular the emission wavelength, can be directly controlled via size and composition in GaN/AlN or AlGaN/AlN quantum dots. Furthermore, a large number of localised states is found already for small GaN/AlN quantum dots. An existing theoretical study for InGaN/GaN quantum dots [251] finds a much smaller number of localised electronic states (only 1 to 3 electron states). The high number of localised states found within this study is explained by the larger conduction and valence band offsets between GaN and AlN as compared to an InGaN/GaN system. This effect is amplified by the fact that the number of localised states within the present work was calculated for pure GaN quantum dots embedded in a pure AlN matrix whereas in Ref. [251] the In composition in InGaN does not exceed 50% and is even more decreasing from the quantum dot center towards the GaN matrix.

The number of edges in a regular-polygon based truncated pyramidal quantum dot is found to have a characteristic influence on degeneracies for energetically close states. However, this parameter is not easily controlled during the growth process whereas some geometries (e.g. triangular, squared, hexagonal or circular bases) are energetically more favourable due to the symmetry of the underlying crystal structure.

4.1.4 Polar vs. nonpolar grown III-nitride quantum dots

The wurtzite phase is the thermodynamically most stable phase for the III-nitride materials. Within this crystal phase, strong built-in potentials occur due to spontaneous and piezoelectric polarisations. III-nitride quantum dots are commonly grown along the polar [0001] direction, where strain and polarisation lead to built-in potentials in the order of MV/cm and therefore to a spatial separation of electron and hole states which results in a reduced recombination rate and reduced light emission efficiency. While ideal nonpolar grown quantum wells having no interfaces oriented along the [0001] direction eliminate this problem, nonpolar grown quantum dots will still cause built-in fields that induce a spatial separation of electrons and holes.

Within the past five years, much progress was made in the understanding of nonpolar quantum dots in experiment. Quantum dot [63] and even quantum wire [6] growth on a-plane ($11\bar{2}0$) and m-plane ($1\bar{1}00$) surfaces have been observed and a strong reduction of the quantum confined Stark effect was found [6, 84] together with strong excitonic localisation [203]. Misfit dislocation densities are found to be smaller than the quantum dot density, making nonpolar quantum dots ideal systems for efficient light emitters [76]. The improved light emission properties in comparison to quantum dots grown on polar surfaces are also suspected to be a result of the strain fields and their piezoelectric contribution to the built-in electrostatic potential [60, 76, 84].

However, only a few theoretical investigations on nonpolar III-nitride quantum dots are available so far. Schulz et al. [212] found the piezoelectric potential in GaN quantum dots to have a dramatic influence on the charge carrier localisation even in nonpolar quantum dots. These studies were based on an effective mass model and performed for the assumption that the geometry is the same for quantum dots oriented either in c-direction or in a-direction. However, realistic quantum dots have a different shape when oriented in the polar or in a nonpolar direction due to the underlying crystal structure. This fact has to be taken into account when comparing these quantum dots. Within this chapter, the charge carrier wave functions and eigenenergies as well as the electron-hole overlap in realistic polar and nonpolar quantum dots will be compared. The employed eight band $\mathbf{k} \cdot \mathbf{p}$ approach furthermore allows to perform systematic studies of various polar and nonpolar quantum dots. The influence of shape and size on the electronic properties can thus be easily investigated, allowing conclusions how the experimentally observed systems can be modified in order to improve their applicability for light emission devices [152]. Furthermore, the strain-induced periodic stacking of quantum dots along growth direction might possibly affect the electronic properties. The distance between stacked quantum dots can be easily modified in experiment by allowing thicker AlN spacer layers on top of the quantum dots. To study the influence of this periodic ordering, the distance between neighbouring quantum dots is varied allowing to understand the influence of periodic images of a nonpolar quantum dot.

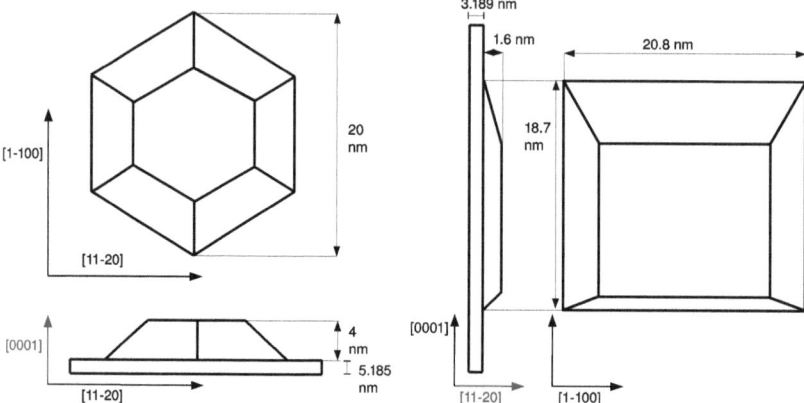

Figure 4.17: *Polar (left) and nonpolar (right) wurtzite GaN quantum dots in AlN as observed in Refs. [77] and [248]. The growth axis is marked red.*

Polar and nonpolar quantum dot models

Fig. 4.17 shows a polar and a nonpolar GaN quantum dot embedded in an AlN matrix as observed by Widmann [248] and Founta [77]. Experimental observations, i.e. HRTEM images, are shown for both systems in Fig. 4.18.

While polar grown quantum dots in wurtzite GaN are observed to have a well defined truncated hexagonal pyramid shape [247], the nonpolar grown quantum dot shows a rather unintuitive geometry, according to Founta et al.:

1. The base of the quantum dot is not quadratic but rectangular. The ratio between the base lengths b_y along $[1\bar{1}00]$ and b_z in $[0001]$-direction is found to be larger than 1, i.e. the dot is elongated from a quadratic base towards the $[1\bar{1}00]$-direction. Typical base lengths for nonpolar GaN/AlN quantum dots range from $a^{[1\bar{1}00]} = 20.8$ nm and $a^{[0001]} = 18.7$ nm to $a^{[1\bar{1}00]} = 25.3$ nm and $a^{[0001]} = 24.6$ nm. The height of the dots ranges from $h^{[11\bar{2}0]} = 1.6$ nm to $h^{[11\bar{2}0]} = 2.6$ nm. With typical base diameters of 17 to 23 nm and heights of 3.5 to 4.5 nm in polar grown quantum dots [248], the dimensions of polar and nonpolar quantum dots are comparable.

2. The side facets oriented in $[0001]$ and $[000\bar{1}]$ direction span up different angles with the wetting layer. The facet at the $[0001]$ side of the dot spans an angle of only 15° between facet and wetting layer, but the angle between the $[000\bar{1}]$ facet and the wetting layer is 50°.

3. Polar GaN/AlN quantum dots are typically randomly distributed on the wetting layer. The nonpolar dots, on the other hand, show a periodical alignment along the $[1\bar{1}00]$-

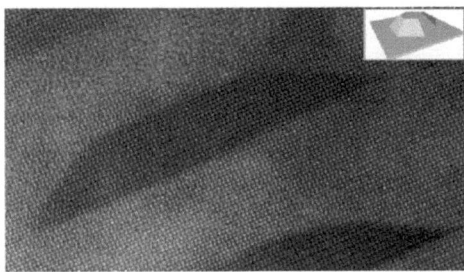 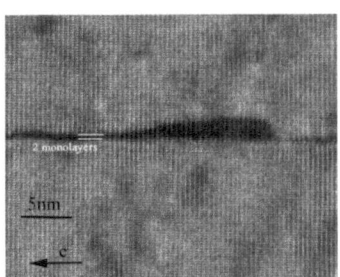

Figure 4.18: *HRTEM images of polar (left, taken from Ref. [248]) and nonpolar (right, from Ref. [6]) GaN/AlN quantum dots. For polar quantum dots, the inlet in the left picture shows a schematic picture of the shape.*

direction. The periodicity length is found to be about 30 nm, as previously reported by Onojima et al. [177].

Polar and nonpolar quantum dots are grown in a superlattice system, where dots are capped by AlN. The distance between two dots in growth direction is about 8 nm for polar and 6 nm for nonpolar quantum dots. In most samples, a strain induced stacking of quantum dots on top of each other along growth direction has been observed for polar as well as for nonpolar quantum dots.

For the calculation of the electronic properties, the geometries depicted in Fig. 4.17 have been used to model the quantum dots, simulating the experimentally observed systems. In both cases, a wetting layer of 2 monolayers has been assumed, according to the observations in Refs. [77, 248].

For the **polar dot**, the cell size was chosen to be $40 \times 40 \times 8$ nm^3 for a dot diameter of 20 nm and a dot height of 4 nm. While the cell size along [0001] is chosen to reflect the stacking of quantum dots in a superlattice, the lateral cell size is sufficiently large to simulate an isolated quantum dot in a random distribution of dots on the wetting layer.

In case of the **nonpolar dot**, the cell size was chosen to be 6 nm along [11$\bar{2}$0], 30 nm along [1$\bar{1}$00] and 40 nm in the [0001] direction. This simulates the experimentally observed system of stacked quantum dots along the growth direction ([11$\bar{2}$0]) and periodically aligned dots along [1$\bar{1}$00]. Along [0001], the dots are considered to be isolated from the neighbouring quantum dots. The base lengths of the dot were chosen as $a^{[1\bar{1}00]} = 20.8$ nm, $a^{[0001]} = 18.7$ nm and the height is $h^{[11\bar{2}0]} = 1.6$ nm.

The parameter set used for the calculation of strain, piezoelectric potentials, charge carrier wave functions and binding energies can be found in Tab. 4.6. For the calculation of the piezoelectric potential V_p, the piezoelectric constant e_{15} is required in systems where shear strains occur, as it is the case in quantum dots (see Eq. (2.47)).

As pointed out by Williams et al. [250], there is a large uncertainty not only for the value

Parameter	GaN	AlN	Parameter	GaN	AlN
a (Å)	3.189	3.112	E_g (eV)	3.24	6.47
c (Å)	5.185	4.982	E_{vb} (eV)	0.8	0.0
C_{11} (GPa)	390	396	Δ_{so} (eV)	0.014	0.019
C_{12} (GPa)	145	137	Δ_{cr} (eV)	0.034	-0.295
C_{13} (GPa)	106	108	$m_e^{\parallel}$ (m_0)	0.186	0.322
C_{33} (GPa)	398	373	$m_e^{\perp}$ (m_0)	0.209	0.329
C_{44} (GPa)	105	116	A_1	-5.947	-3.991
P_{sp} (C/m^2)	-0.034	-0.090	A_2	-0.528	-0.311
e_{33} (C/m^2)	0.67	1.5	A_3	5.414	3.671
e_{31} (C/m^2)	-0.34	-0.53	A_4	-2.512	-1.147
κ	9.6	8.5	A_5	-2.510	-1.329
			A_6	-3.202	-1.952
			$a_c^{\parallel}$	-9.5	-12.0
			$a_c^{\perp}$	-8.20	-5.4
			D_1	-3.00	-3.00
			D_2	3.60	3.60
			D_3	8.82	9.60
			D_4	-4.41	-4.80
			D_5	-4.00	-4.00
			D_6	-5.10	-5.10

Table 4.6: *Material parameters for wurtzite GaN and AlN. Effective masses, the A_i's, E_g and Δ_{cr} are taken from Ref. [202]. P_{sp}, e_{33} and e_{31} are taken from [26]. All other parameters are taken from Ref. [239].*

of e_{15} in GaN and AlN, but also for its sign. Positive values for e_{15} have been reported from experiment [161] and theory [28] for AlN and GaN. However, negative values for e_{15} can be derived as an estimate from the cubic piezoelectric constant e_{14} using the assumption that $e_{15} = e_{31} = -(1/\sqrt{3})e_{14}$ [39]. Additionally, e_{15} was experimentally observed to be negative in AlN by Bu [36] and Tsubouchi [232].

In order to estimate the maximum and the minimum effect on the built-in polarisation in polar and nonpolar quantum dots, the maximum and minimum literature value of e_{15} are considered. However, for the final results, the positive values of e_{15} obtained from ab initio calculations reported in Ref. [28] are employed.

Built-in potentials in polar and nonpolar quantum dots

The built-in potentials for the polar and nonpolar grown quantum dots were calculated by minimising the strain free energy and solving the Poisson equation, according to Sec. 2.3.4.

The piezoelectric potential for the polar grown, hexagonal quantum dot is shown in Fig. 4.19. Here, e_{15} was varied from 0.33 C/m^2 in GaN and 0.42 C/m^2 in AlN (Fig. 4.19 a)

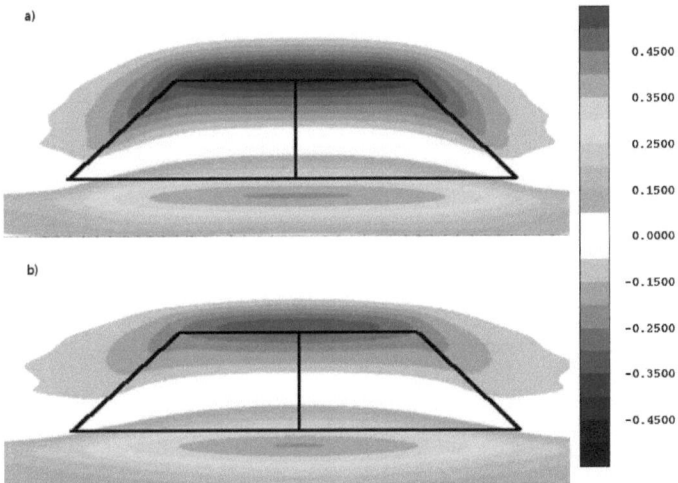

Figure 4.19: *Piezoelectric potential in eV for* $e_{15}(GaN) = 0.33$ C/m^2, $e_{15}(AlN) = 0.42$ C/m^2 *(a) and* $e_{15}(GaN) = -0.48$ C/m^2, $e_{15}(AlN) = -0.6$ C/m^2 *(b)*.

to -0.48 C/m^2 in GaN and -0.6 C/m^2 in AlN (Fig. 4.19 b).

It is clearly visible that different values of e_{15} introduce only slight, quantitative modifications to the piezoelectric potential.

The piezoelectric potential for nonpolar quantum dots is shown in Fig. 4.20. Here, the different values for e_{15} produce significant modifications both quantitatively and qualitatively. It can be seen that the piezoelectric potential resulting from shear strain effects increases or decreases the total polarisation potential, respectively, at the [0001] and the [000$\bar{1}$] edges of the nonpolar dot, depending on the choice of e_{15}. However, the spontaneous polarisation still leads to an attractive potential for electrons at the top and an attractive potential for holes on the bottom of the quantum dot. Additionally, a strong influence of neighbouring quantum dots is seen in the electrostatic potential, resulting from the small distance between the periodically ordered quantum dots. For a quantitative comparison, Fig. 4.21 shows a plot of the polarisation potential along the [0001] direction for the considered values of e_{15}. It is nicely visible, that the polarisation potential in the nonpolar dot is stronger than the one in the polar dot.

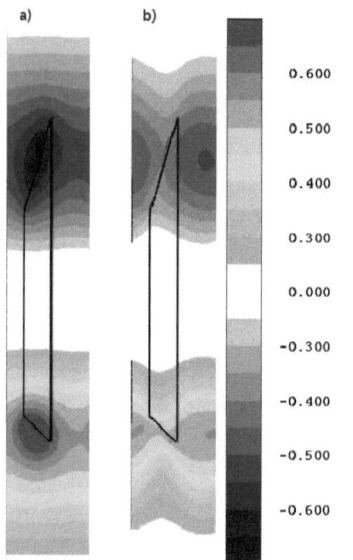

Figure 4.20: *Piezoelectric potential in eV for $e_{15}(GaN) = 0.33$ C/m^2, $e_{15}(AlN) = 0.42$ C/m^2 (a) and $e_{15}(GaN) = -0.48$ C/m^2, $e_{15}(AlN) = -0.6$ C/m^2 (b). For nonpolar dots, the value of e_{15} influences the piezoelectric potential both qualitatively and quantitatively.*

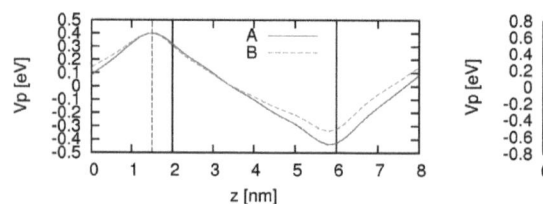
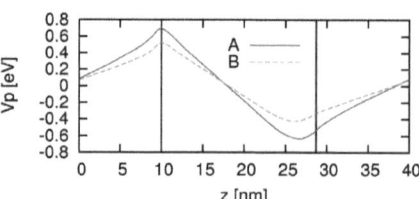

Figure 4.21: *Polarisation potential along the [0001] direction for the polar dot (left) following a line through the dot's center and for the nonpolar dot (right) along an axis through the dot at its base for positive values of e_{15} (A) and negative values (B).*

Localisation of electron and hole states

Using the eight band $\mathbf{k} \cdot \mathbf{p}$ model from Sec. 2.3, the energetically lowest lying electron and hole states have been calculated for the polar and the nonpolar quantum dot. The overlap between electron and hole states has been calculated using Eq. (2.53) in order to estimate the light emission efficiency of such nanostructures.

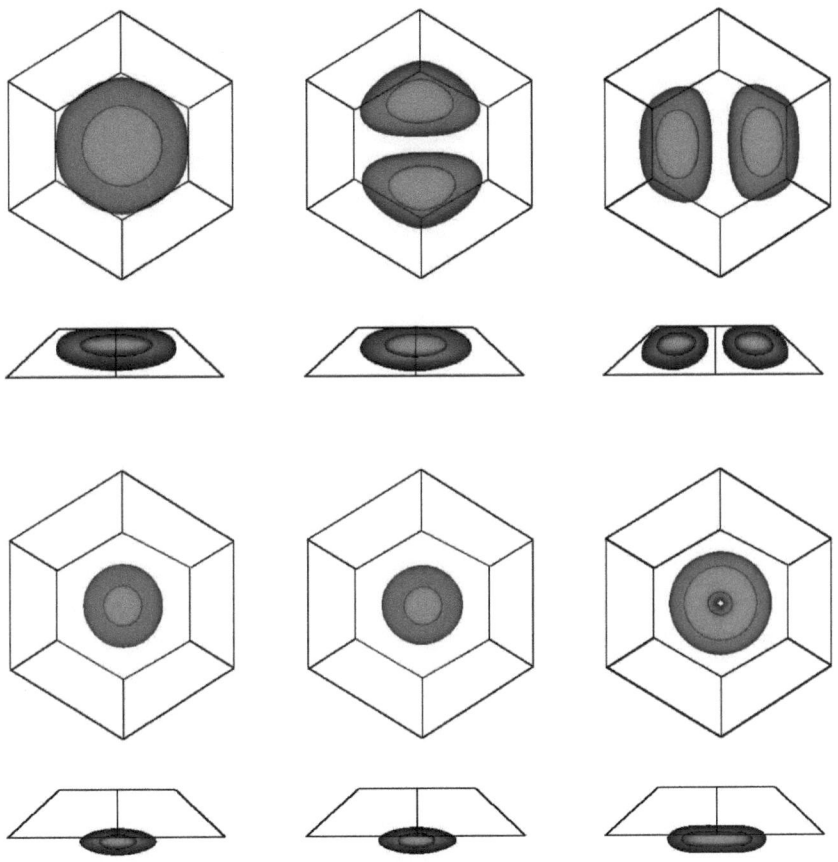

Figure 4.22: *Electron (top) and hole states (bottom) in a polar GaN quantum dot. Probability densities for 90% (red) and 50% (green) are shown from atop and side view. The piezoelectric constant e_{15} is the one from case A.*

The electron and hole charge densities are shown in Fig. 4.22 with the piezoelectric con-

	0.33 C/m²	-0.48 C/m²
e_0	3.9868	4.0401
e_1	4.0327	4.0851
e_2	4.0338	4.0861
h_0	0.9831	0.9682
h_1	0.9759	0.9611
h_2	0.9755	0.9591

Table 4.7: Binding energies in a polar GaN quantum dot for $e_{15} = 0.33$ C/m² and $e_{15} = -0.48$ C/m² in GaN (left). Below: Overlap matrix for electron and hole states for $e_{15} = 0.33$ (below left) and $e_{15} = -0.48$ C/m² (below right).

	ϱ_0^h	ϱ_1^h	ϱ_2^h
ϱ_0^e	$4.9\cdot10^{-11}$	$4.9\cdot10^{-11}$	$4.3\cdot10^{-11}$
ϱ_1^e	$2.6\cdot10^{-11}$	$2.4\cdot10^{-11}$	$3.2\cdot10^{-11}$
ϱ_2^e	$2.5\cdot10^{-11}$	$2.5\cdot10^{-11}$	$3.2\cdot10^{-11}$

	ϱ_0^h	ϱ_1^h	ϱ_2^h
ϱ_0^e	$8.9\cdot10^{-11}$	$8.9\cdot10^{-11}$	$7.9\cdot10^{-11}$
ϱ_1^e	$4.3\cdot10^{-11}$	$4.0\cdot10^{-11}$	$5.6\cdot10^{-11}$
ϱ_2^e	$4.1\cdot10^{-11}$	$4.1\cdot10^{-11}$	$5.5\cdot10^{-11}$

stant e_{15}(GaN) $= 0.33$ C/m². The localisation of electrons at the top and hole states at the bottom of the quantum dot is caused by the piezoelectric potential and was previously observed by many researchers [7, 72, 251]. A different choice of e_{15} does not produce qualitative changes in the shape of the charge densities shown in Fig. 4.22. Quantitative changes, however, can be seen in the binding energies in Tab. 4.7. For $e_{15} = -0.48$ C/m², the overlap between electrons and holes is increased by almost a factor of 2. This is consistent with a stronger localisation of electrons and holes perpendicular to the growth direction. While the distance between electron and hole states does not differ for the applied values of e_{15}, the charge densities show stronger confinement in lateral direction.

The electron and hole states for the nonpolar grown quantum dot are shown in Fig. 4.23. A strong localisation of electrons at the top and holes at the bottom of the quantum dot can be seen. The corresponding binding energies and the overlap matrices for the maximum and minimum chosen value for e_{15} are given in Tab. 4.8. The overlap matrices are five orders of magnitude smaller than those for the polar grown quantum dot showing a very small spatial overlap of electron and hole states at least for the states closest to the conduction and valence bands which govern the excitonic properties. Different values for e_{15} increase the overlap between electron and holes about a factor of 2, as observed also for the polar grown dots. However, as these overlaps are still five orders of magnitude smaller than in the polar dots, different values for e_{15} do not lead to a qualitative increase of the electron-hole overlap.

To investigate the influence of the actual shape of the quantum dot on the charge carrier overlap and thus the light emission efficiency, the above calculations have been performed additionally for the simplified quantum dot structures given in Ref. [212] and employing positive piezoelectric constants e_{15} in GaN and AlN. In the simplified model, rectangular-based, truncated pyramids have been assumed for the quantum dot shape in polar and in nonpolar direction. In particular, the interfaces of the nonpolar quantum dot in the [0001] and the [000$\bar{1}$] direction are assumed to have the same slope. It turns out that the resulting overlap matrix is only slightly different from the overlap obtained for realistic nonpolar

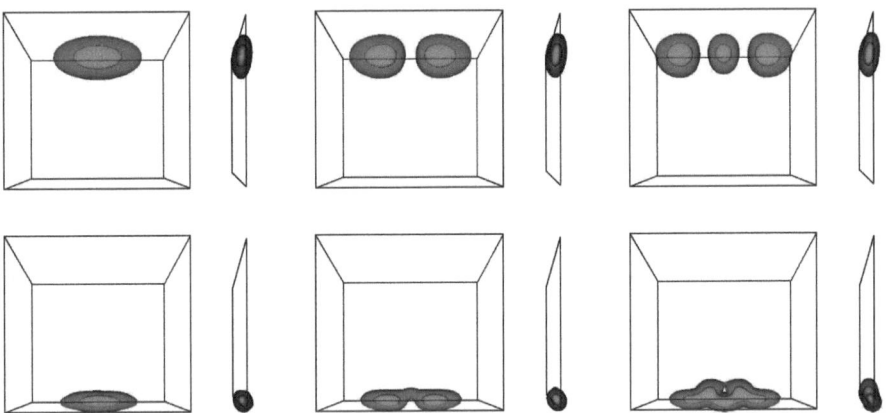

Figure 4.23: *Electron (top) and hole states (bottom) in a nonpolar GaN quantum dot for $e_{15}(GaN) = 0.33$ C/m². Probability densities for 90% (red) and 50% (green) are show from atop $[11\bar{2}0]$ and side view along $[1\bar{1}00]$.*

	0.33 C/m²	-0.48 C/m²
e_0	3.9641	4.1138
e_1	4.0010	4.1437
e_2	4.0418	4.1767
h_0	1.3425	1.1242
h_1	1.3158	1.1029
h_2	1.2945	1.0911

Table 4.8: *Binding energies in a nonpolar GaN quantum dot for $e_{15} = 0.33$ C/m² and $e_{15} = -0.48$ C/m² in GaN (left). Below: Overlap matrix for electron and hole states for $e_{15} = 0.33$ (below left) and $e_{15} = -0.48$ C/m² (below right).*

	ϱ_0^h	ϱ_1^h	ϱ_2^h		ϱ_0^h	ϱ_1^h	ϱ_2^h
ϱ_0^e	5.7·10⁻¹⁵	3.2·10⁻¹⁵	2.4·10⁻¹⁵	ϱ_0^e	3.3·10⁻¹⁵	2.0·10⁻¹⁵	1.7·10⁻¹⁵
ϱ_1^e	3.0·10⁻¹⁵	3.7·10⁻¹⁵	2.1·10⁻¹⁵	ϱ_1^e	1.6·10⁻¹⁵	1.9·10⁻¹⁵	1.3·10⁻¹⁵
ϱ_2^e	2.8·10⁻¹⁵	2.0·10⁻¹⁵	2.0·10⁻¹⁵	ϱ_2^e	1.6·10⁻¹⁵	1.1·10⁻¹⁵	1.00·10⁻¹⁵

quantum dots, e.g. the ground state overlap is $8.6 \cdot 10^{-15}$ in the simplified and $5.5 \cdot 10^{-15}$ in the realistic nonpolar quantum dot. For the polar geometry, where realistic dots have a hexagonal shape and a larger height, the simplified geometry produces a charge carrier overlap of $4.7 \cdot 10^{-10}$, which is one order of magnitude larger than for the realistic polar quantum dot. This is a result of the reduced size of the dot along the [0001] direction, which leads to a stronger confinement due a narrow band offset potential and furthermore to a less intensive spatial separation of electrons and holes induced by the polarisation potential which is smaller in the flat geometry of the simplified model system.

Periodic stacking of nonpolar quantum dots

Nonpolar GaN quantum dots have been observed to be periodically stacked on top of each other in the [11$\bar{2}$0] growth direction. This stacking is induced by strain and the distance between dots periodically ordered along growth direction can be easily controlled in experiment by increasing the AlN spacer layer between two neighbouring GaN layers. This periodicity along growth direction modifies the built-in electrostatic potentials, as can be already seen in Fig. 4.19 where the effect of periodicity is directly visible in the non-zero potentials at the cell boundaries. To study the influence of periodic images on the spatial charge carrier separation, the distance between two neighbouring dots along [11$\bar{2}$0] is systematically varied. In Fig. 4.24, the polarisation potential is plotted along the [0001] direction close to the wetting layer through the nonpolar dot for dot distances from 4.5 to 24 nm along the growth direction. It can be seen that the periodicity of the structure has also no significant influence on the charge carrier localisation. In particular, no qualitative effect on the electron and hole localisation can be expected. Furthermore, only slight quantitative changes occur for distances above 9 nm, i.e., this distance is already a good description for an isolated quantum dot for the built-in electrostatic potential. It can be concluded, that the distance between the periodically stacked quantum dots is no parameter that can be used to effectively increase the electron-hole overlap and thus the efficiency of light emission processes.

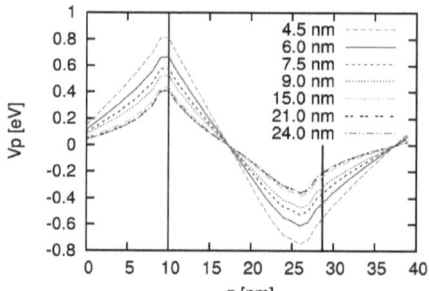

Figure 4.24: *Polarisation potential in a nonpolar dot along the [0001] direction for various dot distances along the growth [11$\bar{2}$0] direction.*

Correspondingly, the charge carrier localisation and thus the electron-hole overlap remains qualitatively unaffected by the periodicity of the system. For a cell size of 6 nm along [11$\bar{2}$0], the overlap between the electron and the hole ground state is found to be $5.5 \cdot 10^{-15}$ whereas for a 12 nm long cell, this overlap element is $2.8 \cdot 10^{-15}$. The interaction with neighbouring quantum dots modifies the electron-hole overlap by a factor of approximately 2, which is again no qualitative improvement of the electron-hole overlap.

Influence of the quantum dot size on the charge carrier localisation

The overlap between electron and hole states is expected to increase for smaller quantum dots. In order to understand the influence of the quantum dot size on the charge carrier overlap, both the polar and the nonpolar quantum dot have been rescaled downwards up to 30% of their original dimensions without modifying the quantum dot's shape. The overlap matrix element for the electron ground state with the first three hole states and for the hole ground state overlapping with the first three electron states as a function of the quantum dot's size is given in Fig. 4.25 for the polar and the nonpolar quantum dot. The charge carrier

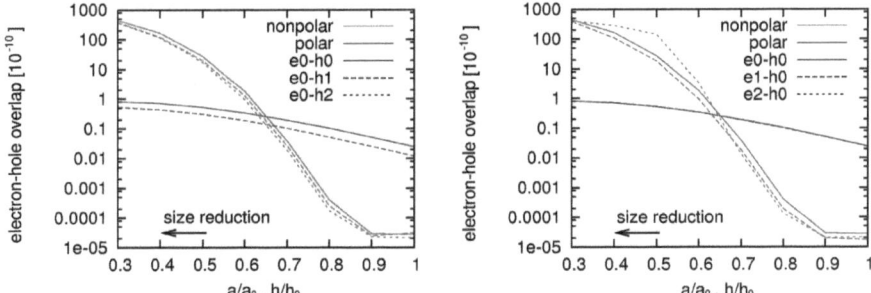

Figure 4.25: *Overlap between electron ground state and first three hole states (left) and hole ground state and first three electron states (right) as a function of the quantum dot's size for the polar (blue) and the nonpolar quantum dot (red). Please note the logarithmic y-axis.*

overlap for the **polar quantum dot** increases linearly with reduced dimensions. For the **nonpolar quantum dot**, however, it can be seen that the charge carrier overlap increases exponentially for smaller dot dimensions. Moreover, the overlap element between the hole ground state Ψ_0^h and the second excited electron state Ψ_2^e increases abruptly by two orders of magnitude when resizing the quantum dot from 70% to 60% of its original base lengths and height. This large increase is explained by a qualitative change in the electron wave function. While the second excited electron state is d-like and oriented along $[1\bar{1}00]$ for quantum dot sizes above 60% of the original size, this state is found to be a p-like, $[0001]$-oriented state for quantum dot sizes of 60% and less of the original size. The orientation of this orbital leads to a much larger spatial overlap with the hole ground state (see. Fig. 4.26). Correspondingly, the overlap between second excited electron and hole ground state becomes larger than the overlap of the hole with the electron ground state (Tab. 4.9).

The binding energies of the three electron and hole states closest to the band gap are shown as a function of the quantum dot size in Fig. 4.27. It is nicely to see that for smaller dimensions, the energy of the electrons and holes moves away from the band gap and, additionally, the energy difference between energetically neighbouring states increases. This effect is a result of the increased quantum confinement in smaller structures. Additionally,

	ϱ_0^h	ϱ_1^h	ϱ_2^h
ϱ_0^e	$3.7 \cdot 10^{-10}$	$2.6 \cdot 10^{-10}$	$1.9 \cdot 10^{-10}$
ϱ_1^e	$1.9 \cdot 10^{-10}$	$3.6 \cdot 10^{-10}$	$3.6 \cdot 10^{-10}$
ϱ_2^e	$6.7 \cdot 10^{-10}$	$10.8 \cdot 10^{-10}$	$11.8 \cdot 10^{-10}$

Table 4.9: *Charge carrier overlap in a nonpolar quantum dot with 60% of its original dimensions.*

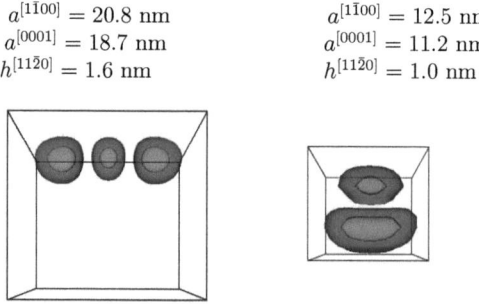

$a^{[1\bar{1}00]} = 20.8$ nm
$a^{[0001]} = 18.7$ nm
$h^{[11\bar{2}0]} = 1.6$ nm

$a^{[1\bar{1}00]} = 12.5$ nm
$a^{[0001]} = 11.2$ nm
$h^{[11\bar{2}0]} = 1.0$ nm

Figure 4.26: *Second excited electron state Ψ_2^e for the original quantum dot (left) and for a quantum dot size of 60% of the original base lengths and height (right).*

Fig. 4.28 shows the energy difference between the three first excited electron states and the electron ground state. This plot makes the crossing of the binding energies of the third and the second excited state visible for a quantum dot size of approx. 54% of the original dimensions, identifying this value as the critical size for a change of a d-like, [1$\bar{1}$00]-oriented state to a p-like [0001]-oriented state.

Conclusions: Polar versus nonpolar grown GaN quantum dots

In this section, we have addressed the question if the emission efficiency of wurtzite GaN quantum dots can be improved by employing a nonpolar growth process. For this purpose, the key electronic properties of polar and nonpolar quantum dots were compared. The critical physical influence opposing a better light emission is the built-in electrostatic potential which induces a spatial separation and thus a weak recombination rate of electrons and holes. The influence of these potentials is found to spatially separate electrons and holes even more in nonpolar grown systems, compared to polar ones. This result is in qualitative agreement with the recent result of Ref. [212], but quantitatively much higher for realistic quantum dot structures considered within the present work. However, the experimentally found overlap [75] is much larger than what is predicted within this work. This could be

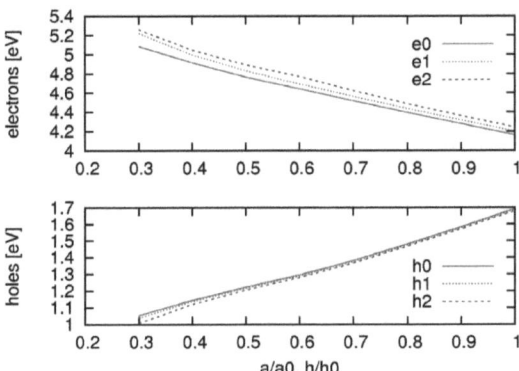

Figure 4.27: *Binding energies of the three electron and hole states closest to the band gap as a function of the quantum dot size.*

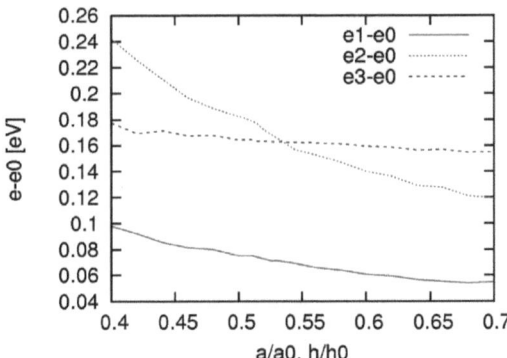

Figure 4.28: *Difference between first three excited electron states and the ground state as a function of quantum dot size.*

caused by so-called skew excitons, where electrons recombine with higher excited hole states. Due to the higher energy of these hole states, a localisation that causes a stronger overlap with the electron states is possible. It is also possible that the experimentally observed overlap is larger than what is predicted within this work due to doping in the vicinity of the quantum dot, leading to modifications of the band edges. Nevertheless, the present work identifies the key parameters which allow to systematically increase the overlap between electrons and holes and, additionally, points out which parameters have only minor influence.

The effect of the piezoelectric constant e_{15} has been studied for the two extremum values available in literature, namely for -0.48 and 0.33 C/m^2 in GaN. This parameter clearly influences the electron and hole binding energies, but does not lead to a significantly better overlap of electron and hole states in the case of nonpolar grown GaN quantum dots.

Furthermore, the influence of strain-induced periodic stacking of nonpolar quantum dots along growth direction was investigated. However, it was found that neighbouring periodic images have no significant influence on the charge carrier localisation.

Calculations with the model geometries of Ref. [212] have been performed to investigate the influence of the quantum dot's shape. For the nonpolar dot geometry, the more realistic shape in Fig. 4.17 was found to induce only slight modifications of electron and hole overlap, whereas the charge carrier overlap in the polar geometry in Ref. [212] yields an overlap which is larger than in realistic polar geometries by a factor of 10. This result stems mainly from the height along the [0001] direction, which is larger in the realistic system by a factor of 2, compared to the model geometry employed by Schulz et al..

Smaller dot dimensions can dramatically increase the electron-hole overlap. In particular, for the studied nonpolar quantum dot, a qualitative change of the second excited electron state was found when decreasing the quantum dot's dimensions to 54% of the original ones. This qualitative change leads to an abrupt increase of the electron-hole overlap of two orders of magnitude. It is quite certain, that similar effects occur for higher excited states, in particular for the hole states that localise in a much larger number than the electron states. However, these states have not been investigated here since their contribution to light emission is small.

In general, the increase of the electron-hole overlap in nonpolar GaN quantum dots is much larger than in polar quantum dots with reduced size. This is explained by the orientation of the charge carriers. In polar quantum dots, electrons and holes are expanded in a plane parallel to the growth plane due to the flat shape of the quantum dot. Due to the polarisation potential, the electrons are located at the top and the holes at the bottom of the dot. By decreasing the dimensions of the polar quantum dot, the electron and hole states oriented parallel towards each other are pushed together, resulting in a weak increase of the charge carrier overlap. In nonpolar quantum dots, electrons and holes are both localised in the same plane, spanned by the [0001] and the [1$\bar{1}$00] direction. Due to the shape of the quantum dot, a strong confinement is already present in the [11$\bar{2}$0] direction. Again, electrons locate at the top (the [0001]-edge) and holes at the bottom [000$\bar{1}$] edge. Reducing the size of the nonpolar quantum dot pushes the electron and hole states located in the same plane together. The resulting increase of the overlap is much larger than in polar quantum dots. It is therefore concluded, that reducing the size of nonpolar GaN/AlN quantum dots is the most promising approach to improved light emission efficiency.

4.2 Quantum wires and dislocations

Semiconductor quantum wires have attracted much research interest in the past years [133, 242] due to their broad application spectrum ranging from transistors [61, 231] and nano-logic gates [108] to photo detectors [243] and light emission devices [117, 215]. A charge carrier confinement along two dimensions is expected to lead to a reduced threshold current required to activate light emission in comparison to quantum well based light emitters. Further, it reduces the spectral line width [52], which enables the fabrication of devices that require coherent light emission with low energy consumption. A main advantage of quantum wires in comparison to quantum dots is the much higher ability to control diameter and position of nanowires during the growth process [103] and thus the emission wavelength.

The electronic structure of quantum wires, and thus the charge carrier localisation, can be simulated as a two-dimensional problem while the charge carriers are free to move along the third dimension. Such systems are commonly referred to as one-dimensional nanostructures.

In a quantum wire system, charge carrier localisation results from the conduction band and valence band offset of the involved materials. However, dislocations and line defects can show comparable localisation effects along two dimensions even in bulk material, where the band edges are constant and can therefore be treated similarly to quantum wires. In particular, formation processes of dislocations [114, 140, 141] and their influence on the electronic properties of materials and, correspondingly, the emission character of specific devices are matter of current research [157, 204, 218]. Strain fields around screw or edge dislocations are suspected to induce a charge carrier localisation and thus to act as unwanted recombination centers in bulk semiconductors.

This chapter provides investigations of quantum-wire like semiconductor nanostructures. The influence of screw dislocations in bulk GaN is studied to understand the charge carrier recombination which occurs around such defects. Furthermore, a systematic study of GaN quantum wires in vacuum has been performed in order to identify and characterise quantum wire structures for applications in novel light emitting devices.

4.2.1 Charge carrier localisation around screw dislocations in bulk GaN

Edge and screw dislocations have been experimentally observed to act as nonradiative recombination centers in GaN. For edge dislocations, it has been shown that even in the case of fully coordinated core atoms (i.e. in the absence of broken bonds) a recombination of electrons and holes around the dislocation occurs [147]. This was found to be a result of local strain fields caused by the edge dislocation. In a similar manner, screw dislocations cause non-zero shear strains. In wurtzite GaN, these shear strains influence the valence band structure, as can be seen in Eq. (A.1) in the Appendix. A localisation of the hole states is expected due to the displacement of the atoms, i.e. the shear strain, around the dislocation. This localisation is suspected to induce unwanted local excitonic recombination effects and therefore reduce the light emission efficiency of GaN-based devices. Correspondingly, a quantitative and qualitative understanding of the influence of shear dislocations on the charge carrier localisation is required. The $\mathbf{k} \cdot \mathbf{p}$ model allows to perform such systematic studies of the influence of strain on the localisation of charge carriers around a screw dislocation employing analytically derived shear strains and well known material properties of the bulk crystal lattice.

Figure 4.29 illustrates a screw dislocation within a continuum picture. A quantitative description of such a lattice failure is the **Burgers vector**. This vector describes the difference between a displaced and the corresponding unperturbed crystal lattice site in magnitude and direction [40]. For a screw dislocation, the Burgers vector **b**, depicted in blue in Fig. 4.29, is parallel to the dislocation axis.

Analytical expressions for the shear strain components

The strain components $\epsilon_{xz}(\mathbf{r})$ and $\epsilon_{yz}(\mathbf{r})$, resulting from a screw dislocation, can be explained as follows: If the z-axis is defined parallel to the Burgers vector, volume elements will not be displaced in x- or y-direction, making $u_x(\mathbf{r}) = u_y(\mathbf{r}) = 0$. In z-direction, the displacement depends on the angle ϕ and increases from 0 for $\phi = 0$ to $b = a_z$ for $\phi = 2\pi$, where a_z is the lattice constant in z-direction:

$$u_z(\mathbf{r}) = \frac{a_z \cdot \phi}{2\pi} = \frac{a_z}{2\pi} \arctan \frac{r_y}{r_x}. \tag{4.1}$$

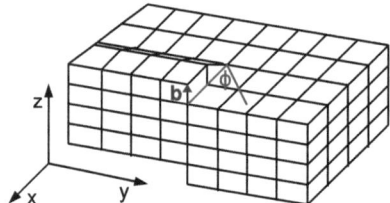

Figure 4.29: *Screw dislocation in a continuum picture. The displacement of each volume element is a function of the angle ϕ. The Burgers vector* **b** *is marked blue.*

The Burgers vector, therefore, is $\mathbf{b} = (0, 0, a_z)$. Now, the strain fields can be calculated from

$$\epsilon_{ij}(\mathbf{r}) = \frac{1}{2}\left(\frac{\partial u_i(\mathbf{r})}{\partial r_j} + \frac{\partial u_j(\mathbf{r})}{\partial r_i}\right). \tag{4.2}$$

With $u_x(\mathbf{r}) = u_y(\mathbf{r}) = 0$ and $u_z(\mathbf{r})$ being constant along the z-direction, it follows that $\epsilon_{xx}(\mathbf{r}) = \epsilon_{yy}(\mathbf{r}) = \epsilon_{zz}(\mathbf{r}) = 0$. Additionally $\epsilon_{xy}(\mathbf{r})$ and $\epsilon_{yx}(\mathbf{r})$ vanish, resulting from zero displacements u_x and u_y. The only emerging non-zero strain components are:

$$\epsilon_{xz}(\mathbf{r}) = \epsilon_{zx}(\mathbf{r}) = -\frac{a_z}{4\pi} \cdot \frac{r_y}{r_x^2 + r_y^2} = -\frac{a_z}{4\pi} \cdot \frac{\sin\phi}{\varrho} \quad \text{and} \tag{4.3}$$

$$\epsilon_{yz}(\mathbf{r}) = \epsilon_{zy}(\mathbf{r}) = -\frac{a_z}{4\pi} \cdot \frac{r_x}{r_x^2 + r_y^2} = \frac{a_z}{4\pi} \cdot \frac{\cos\phi}{\varrho}, \tag{4.4}$$

where $\varrho = \sqrt{r_x^2 + r_y^2}$ is the distance from the screw dislocation and $r_x = r_y = 0$ is the dislocation center. Due to the continuum-like nature of this analytical model, the description of the dislocation core region becomes problematic. In particular, the strain components in Eqs. (4.3) and (4.4) diverge for $\varrho \longrightarrow 0$. This is an unphysical consequence of the continuum picture that does not occur in reality where the center atoms of the screw dislocation are considered to remain undisplaced. Since the $\mathbf{k} \cdot \mathbf{p}$ model itself is a continuum approach, the accuracy of the electronic structure calculations at the dislocation core is at least questionable. Within the present work, this central point is assumed to be an unstrained atom, leaving the bulk band offsets in the core radius R_c around the dislocation center unmodified. This means that no localisation of charge carriers is expected in this core region around the dislocation center. In order to understand a charge carrier localisation around a screw dislocation, however, the strain fields in the area outside the core region are of interest. Furthermore, the continuum elasticity model applied here is valid only for small strains. Therefore, the analytical expressions for $\epsilon_{xz}(\mathbf{r})$ and $\epsilon_{yz}(\mathbf{r})$ are truncated down to zero at the dislocation core radius R_c. The choice of this radius is a bit arbitrary and related to the range of validity of a continuum elasticity description. In order to determine the core radius, the elastic energy stored in a cylinder with the radius R around the dislocation center is calculated via

$$F = A\ln(R/R_c) + E_{\text{core}}, \quad \text{where} \quad A = \mu\frac{\mathbf{b}^2}{4\pi}(1 - \nu) \tag{4.5}$$

is a R-independent prelogarithmic factor [23, 118]. Here, μ is the shear modulus, ν is the Poisson ratio and E_{core} includes the elastic energy contributions from the core region. The total elastic energy is evaluated as a function of $\ln(R)$. The core radius R_c is the value of R above which the function $F(\ln(R))$ starts to behave linear. Ref. [23] estimates this radius for edge dislocations in GaN to be about 0.86 nm. Within the present work, this value has been assumed to be the core radius of the screw dislocation. Employing this core radius limits the occuring strains below a value of 0.1, which is commonly considered to be the limit of a continuum elasticity model. Moreover, it has been systematically checked that slight variations of the core radius do not alter the main messages of this study.

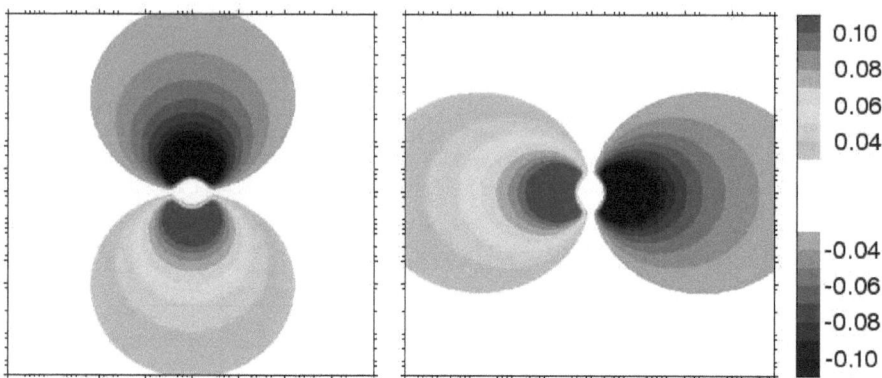

Figure 4.30: *Shear strain components $\epsilon_{xz}(\mathbf{r})$ (left) and $\epsilon_{yz}(\mathbf{r})$ (right) in a cell of 20×20 nm^2 along x and y.*

The valence-band contribution in the $\mathbf{k} \cdot \mathbf{p}$ Hamiltonian for wurtzite structures (see Appendix, Eq. (A.2)) involves the off-diagonal strain components ϵ_{xz} and ϵ_{yz}. These strain tensor components modify the valence-band part of the Hamiltonian via the operator h in Eq. (A.2). The resulting modification of the valence band around the dislocation is expected to yield a certain localisation of hole states, whereas the electron states are not influenced by the strain fields around the dislocation since the conduction-band related part of the $\mathbf{k} \cdot \mathbf{p}$ Hamiltonian does not contain any shear strain contributions. Figure 4.30 shows the strain tensor components $\epsilon_{xz}(\mathbf{r})$ and $\epsilon_{yz}(\mathbf{r})$ in a cell of 20×20 nm^2. The x-axis can be chosen arbitrarily perpendicular to the Burgers vector within this continuum picture. The absolute strains ϵ_{xz} and ϵ_{yz} are mirror symmetric via the x and the y axis. It can be seen that the strain fields reach large values of up to 0.1. For comparison, absolute strains around GaN quantum dots as treated in the previous chapter are commonly in the range below 0.03.

Hole state localisation due to shear strains

The eight-band $\mathbf{k} \cdot \mathbf{p}$-model introduced in Sec. 2.3 allows a fast and systematic study of charge carrier localisation effects induced by shear strains around screw dislocations. Using this approach, the electronic structure of the screw dislocation has been computed. While for the electron states no localisation was observed, a clear localisation of hole states around the dislocation center has been found, resulting from the large shear strain values. The six hole states energetically closest to the valence band edge are shown in Fig. 4.31. It is visible that these states are strongly bound to the dislocation. While only the energetically lowest hole states are shown in Fig. 4.31, the absolute number of hole states localised around the dislocation is much larger. For a core radius of 0.86, 60 localised states were found, being pairwise degenerate due to time reversal symmetry. This strong localisation is not significantly affected by the radius chosen for the truncation of the analytical strains.

Summary: Screw dislocations in GaN

The influence of shear strain around a screw dislocation in bulk wurtzite GaN was investigated. The hole states are found to be strongly bound to the dislocation due to this strain. The non-zero strain contributions $\epsilon_{xz}(\mathbf{r})$ and $\epsilon_{yz}(\mathbf{r})$ were calculated analytically using a continuum elasticity model. Within the analytical model employed in this work, the closer core region with a radius of 0.86 nm around the dislocation center cannot be properly described. Therefore, the strain in this region was artificially set to zero. Nevertheless, the strain contributions in a distance of more than 0.86 nm to the center were found to be sufficiently large to produce a strong coupling of hole states around the dislocation center. Since no localisation of electrons was observed elsewhere, these hole state localisation effects will lead to an increased recombination of electrons and holes in the area of the dislocation. These results are in agreement with recent experimental investigations which find screw dislocations to act as non-radiative recombination centers [5, 59, 96, 104]. This study underlines the importance of a high crystal quality required for the production of light emitting devices. In particular, screw dislocations forming in bulk GaN will continue in other materials that are epitaxially grown on top of the GaN, e.g. InGaN layers, and can thus also reduce the light emission efficiency of quantum-well based devices. Moreover, shear dislocations which form in bulk substrates may continue in quantum dots or wires and modify the emission spectra of these nanostructures. While the present study provides a qualitative explanation of the experimentally observed non-radiative recombination around screw dislocations, more accurate atomistic calculations may allow a more quantitative description of such systems including effects from the core region in future studies.

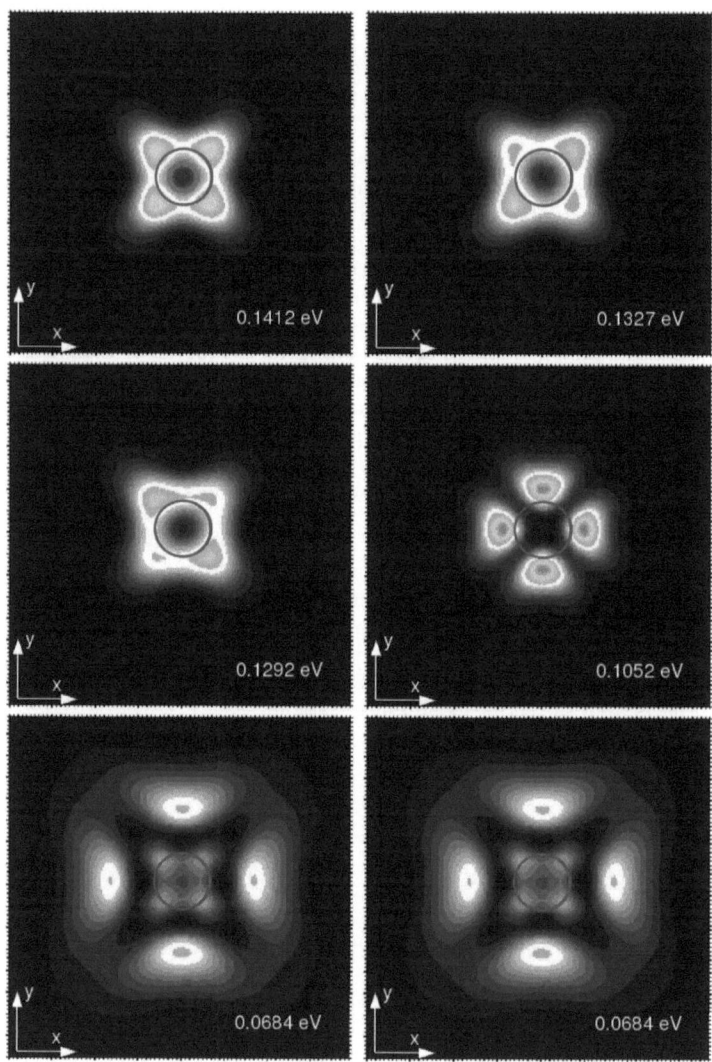

Figure 4.31: *The lowest six hole states (left to right, top to bottom) around a screw dislocation in bulk wurtzite GaN. All states are twofold degenerate. The cell size along x and y-direction is 10×10 nm^2 in this plot. For the calculation, a cell of 20×20 nm has been employed. Green, yellow and brown contours represent high charge densities. The binding energies are given for each state with respect to the bulk GaN valence band edge $E_{\mathrm{vb}} = 0$. The core radius R_c is marked red.*

4.2.2 Wurtzite GaN quantum wires in vacuum

As indicated above, nanowires are highly promising structures for the application in active nanophotonic devices [68, 100, 139, 193]. GaN quantum wires are subject of many recent experimental publications [85, 103, 134, 246]. However, GaN wires reported so far having diameters of 25-100 nm [134] and lengths in the order of μm, are too large to exhibit energy quantisation effects. Although the excitons are localised along two dimensions, the electronic properties for wires of such characteristic dimensions are expected to be purely bulk properties.

In order to achieve a better understanding of energy quantisation in quantum wires and thus to allow for a possible tuning of the light emission spectra of GaN nanowires, the influence of the quantum wire diameter on the charge carrier binding energies is investigated in this section.

Model system and applied formalism

The reference quantum wire was chosen to have a hexagonally shaped geometry with diameters from 4 to 36 nm, grown in [0001] direction and assuming an infinite length. The shape of these model wires reflects experimental observations [134]. In order to study the influence of possible quantum-confinement effects, the diameter ranges also below the experimentally observed scales. The cell size is assumed to be 5 times the quantum wire diameter in directions perpendicular to the growth axis and the GaN wire is surrounded by vacuum. The system is depicted in Fig. 4.32.

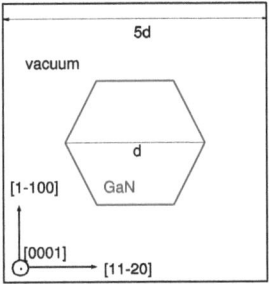

Figure 4.32: *GaN quantum wire geometry (not full-scale).*

The eight band $\mathbf{k}\cdot\mathbf{p}$ model has proven to be a highly efficient tool for various investigations of semiconductor nanostructures within the previous sections. It is therefore also used in the present study to compute the binding energies of charge carriers localised in a GaN quantum wire. Strain contributions do not enter the calculation since the wires are surrounded by vacuum and therefore free to relax. The parameter set for GaN is the same as used in the previous studies of GaN quantum dots and shown in Tab. 4.6. The conduction band offset between GaN and the vacuum is directly described by the electron affinity in GaN. Here, a value of 3.1 eV was used [55]. In order to move an electron from the valence band to the vacuum region, the band gap and the electron affinity have to be overcome, resulting in a valence band offset of $\Delta E_{\mathrm{vb}} = -6.34$ eV from GaN to vacuum. The vacuum effective masses and A_i's were, for simplicity, assumed to be the corresponding GaN values. However, no visible influence of these parameters on the charge carrier binding energies was observed due to the strong localisation of the wave functions inside the quantum wire resulting from the large band offset between the GaN wire and the vacuum matrix.

Results and discussion

The charge density of the first four electron and hole states in a hexagonal quantum wire of 4 nm diameter is shown in Fig. 4.33. It can be seen that typical nodal structures for a two-dimensional potential form, i.e., the ground states of electrons and holes are formed s-like, followed by two p-like states and a d-like third excited state. This behaviour does not qualitatively change for quantum wires with larger diameters. Furthermore, a stronger localisation of the hole states in comparison to the electrons is seen for all studied wire diameters, e.g. in the spatial dispersion of the corresponding ground states.

Figure 4.34 shows the binding energies of the four lowest electron and hole states as a function of the quantum wire diameter. It can be nicely seen that quantisation effects emerge for diameters below approx. 10 nm. This is the edge of what has already been achieved in recent experiments and allows to conclude that only minor improvements of the present growth processes towards smaller diameters are necessary to allow employment of energy quantisation for tuning the electro-optical properties of GaN nanowires.

While in the present study pure GaN wires were investigated, thin $Al_{0.2}Ga_{0.8}N$ coatings as observed by Lari et al. [135] are expected to only slightly increase the localisation and corresponding quantisation effects, in particular by a minimal decrease of the diameter of the pure GaN part of the wire. In conclusion, it is necessary to further decrease the diameter of GaN quantum wires in order to induce significant quantisation effects and thus to control the emission wavelength of these nanostructures.

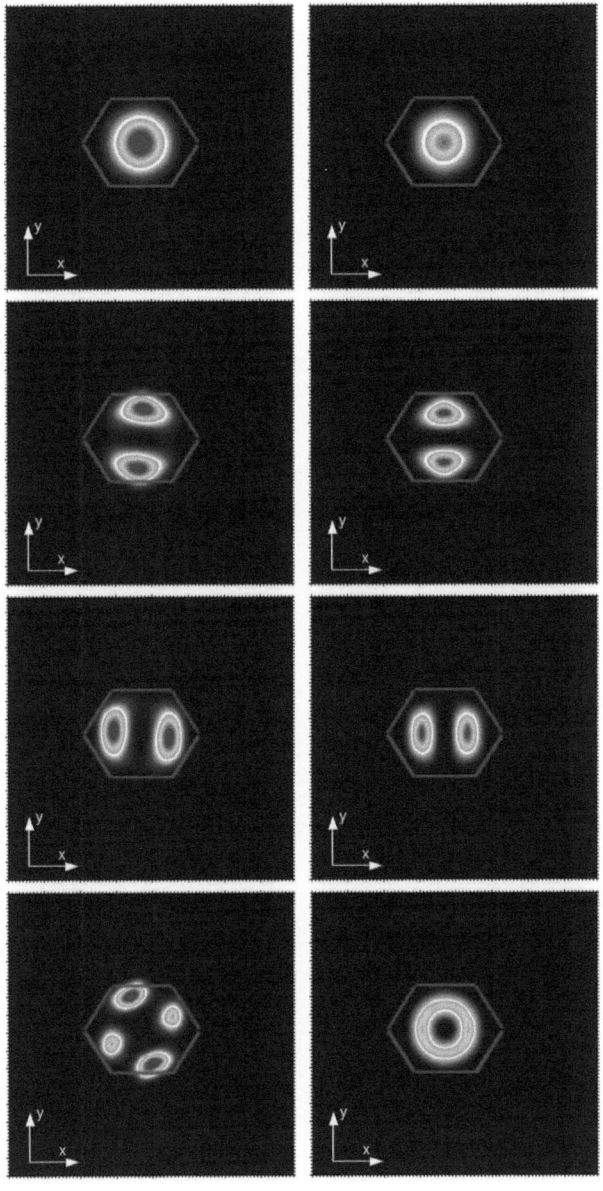

Figure 4.33: *The charge densities of the four energetically lowest electron (left) and hole (right) states in a cell of 10 × 10 nm. Red: wire geometry.*

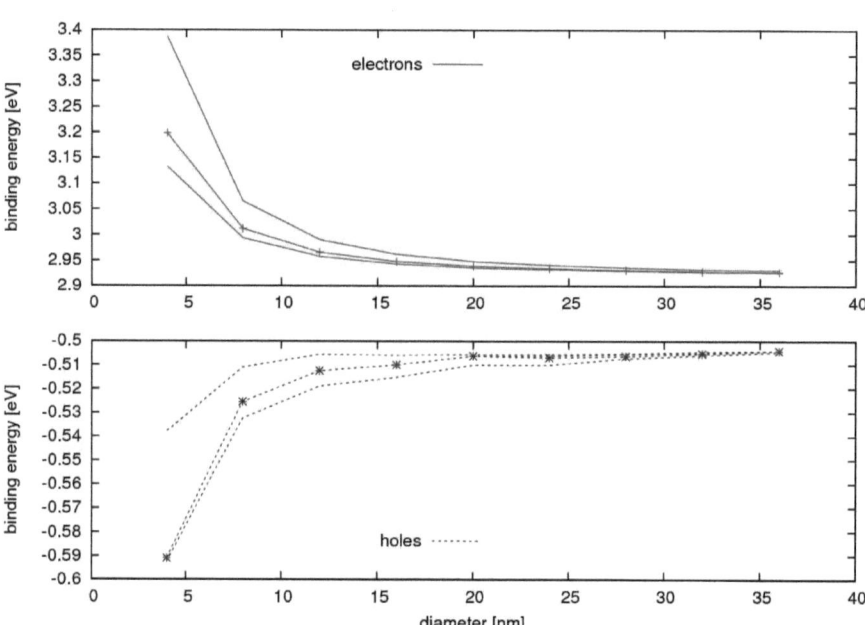

Figure 4.34: *Four energetically lowest electron and hole binding energies as a function of the wire diameter. Lines with crosses represent two degenerate states.*

4.3 Investigations on InGaN quantum wells and superlattices

III-nitride based quantum wells are an excellent basis for the design of light emitting diodes in a spectrum ranging from infrared and the visible spectrum [3, 167, 188] towards the ultraviolet [162, 163] by varying the In content. $Al_{1-x}Ga_xN$ lasers have been applied in the past to achieve light emission in the ultraviolet region [229, 258]. The growth processes of AlGaN or GaN wells on AlN surfaces are well established. A light emission with larger wavelengths can be achieved by employing InN or InGaN wells or superlattices in a GaN matrix. However, the lattice mismatch between bulk InN and GaN is much higher than the one between GaN and AlN, which makes the epitaxial growth of pure InN on a bulk GaN substrate impossible. Therefore, typical $In_xGa_{1-x}N$ quantum wells have In contents ranging from 2% [129] to 30% [158]. Quantum-well based diodes basically consist of a set of potential barriers along one direction. The charge carriers are free to move in two dimensions and are localised along the third dimension. An ideal system can therefore be simulated efficiently as a one dimensional problem within a $\mathbf{k} \cdot \mathbf{p}$ model. A common quantum-well based device for light emission purposes is depicted in Fig. 4.35.

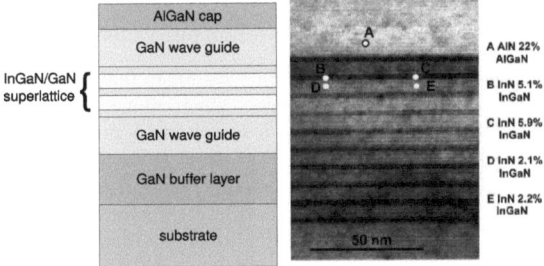

Figure 4.35: *Left: Light emitter based on a III-nitride superlattice. The active region consists of a set of InGaN quantum wells embedded in GaN. Right: HRTEM image of InGaN quantum wells, taken from Ref. [47].*

The elastic and electronic properties of ideal quantum wells in a continuum picture can be described straightforwardly from analytic considerations. However, realistic systems exhibit some deviations from the ideal case which make the determination of elastic and electronic properties more challenging. Inhomogeneous In concentration in InGaN as well as fluctuations of the quantum well thickness occur [47]. So far, the impact on the electronic properties is not well understood.

Within this chapter, polar and nonpolar III-nitride quantum well systems are studied and compared. Special interest is paid to the impact of built-in electrostatic potentials on charge carrier separation and the corresponding decrease of light emission efficiency.

4.3.1 Polar InGaN/GaN superlattices

InGaN/GaN superlattice systems grown on polar surfaces have been used in the past in lasers [165, 166] and light emitting diodes (LEDs) [164]. The main challenge of such devices is the large lattice mismatch between InN and GaN which makes epitaxial growth processes of In-rich InGaN on GaN surfaces problematic. Furthermore, large polarisation effects in wurtzite III-nitrides lead to a spatial separation of electrons and holes and thus to poor recombination rates in polar InGaN/GaN quantum wells.

Within this section, the influence of In content and the InGaN layer thickness on electron and hole binding energies and charge carrier overlap is investigated. This investigation provides a qualitative and quantitative understanding of the influence of these parameters on the electronic properties, allowing to point out ways to improve the efficiency of polar InGaN quantum-well based light emission devices. Furthermore, these studies help to provide a better understanding of the nonpolar grown quantum well systems studied within the next section.

The eight band $\mathbf{k} \cdot \mathbf{p}$ formalism allows an efficient and straightforward analysis of these properties for a wide range of possible systems. Light emission wavelengths are estimated from the electron and hole ground state binding energies.

Strain and polarisation in polar InGaN quantum wells

An ideal semiconductor quantum well epitaxially grown in polar direction and embedded in a matrix material is a one-dimensional problem, which allows to calculate many properties of such a system analytically. This approach is suited to provide a basic understanding of ideal quantum wells before investigating fluctuated InGaN/GaN quantum wells within the following section. In particular, strain fields and the resulting polarisation potential, which both enter the $\mathbf{k} \cdot \mathbf{p}$ Hamiltonian to calculate the electronic properties, are derived analytically within this section.

The **strain fields** can be computed from the matrix and layer material bulk lattice constants, denoted as a_M^0 and a_L^0 in the in-plane and c_M^0 and c_L^0 in the growth direction. The matrix material is assumed to be pure GaN: $a_M^0 = a(GaN)$, $c_M^0 = c(GaN)$, whereas the layer material is the ternary alloy $In_xGa_{1-x}N$ where the bulk lattice constant is linearly interpolated between the GaN and the InN lattice constants:

$$a_L^0 = x \cdot a(InN) + (1-x) \cdot a(GaN) \quad \text{and} \quad c_L^0 = x \cdot c(InN) + (1-x) \cdot c(GaN). \quad (4.6)$$

In systems with layer thicknesses much smaller than the matrix material, the matrix material remains unstrained and the lattice constants in the matrix material are exactly the bulk lattice constants: $a_M = a_M^0$, $c_M = c_M^0$. In the InGaN layer, the in-plane lattice constants match the matrix lattice constants due to the epitaxial growth process: $a_L = a_M^0$. The resulting strain fields ϵ_{xx} and ϵ_{yy} inside the InGaN layer are therefore:

$$\epsilon_{xx} = \epsilon_{yy} = \frac{a_L}{a_L^0} - 1 = \frac{a(GaN)}{a(InGaN)} - 1. \quad (4.7)$$

Minimising the elastic energy in Eq. (2.43) and thus solving Eq. (2.44) yields the following relationship between ϵ_{xx}, ϵ_{yy} and ϵ_{zz}:

$$\epsilon_{zz} = -\frac{C_{13}}{C_{33}}(\epsilon_{xx} + \epsilon_{yy}) = -2\frac{C_{13}}{C_{33}}\epsilon_{xx}. \tag{4.8}$$

Here, the elastic constants $C_{13} = C_{13}(InGaN)$ and $C_{33} = C_{33}(InGaN)$ are again linear interpolations between the bulk InN and GaN elastic constants. In the range of $0 \leq x \leq 0.2$ and for the elastic constants of InN and GaN, ϵ_{zz} is an almost linear function with only weak bowing. Since no shear processes occur in this one-dimensional model, all shear strains ϵ_{ij} with $i \neq j$ are zero.

With the knowledge of all involved strain fields, the polarisation $\mathbf{P}(\mathbf{r})$ can be derived. Following Eq. (2.47), the piezoelectric polarisation in absence of shear strain is given as

$$P_x = P_y = 0 \quad \text{and} \quad P_z = e_{31}(\epsilon_{xx} + \epsilon_{yy}) + e_{33}\epsilon_{zz} + P_{\text{spont}}(InGaN) \tag{4.9}$$

in InGaN and due to the completely strain free growth of the matrix as

$$P_x = P_y = 0 \quad \text{and} \quad P_z = P_{\text{spont}}(GaN) \tag{4.10}$$

in the surrounding GaN. The piezoelectric constants $e_{13} = e_{13}(InGaN)$ and $e_{33} = e_{33}(InGaN)$ are again linear interpolations of the bulk InN and GaN piezoelectric constants. The spontaneous polarisation in InGaN is calculated using a quadratic interpolation (see Eq. (4.18)). However, the bowing of the InGaN polarisation as a function of In content is again rather weak. With the polarisation components P_x and P_y being zero and P_z being two constant values in the InGaN layer and the GaN matrix, the Poisson equation (3.17) can be simplified to:

$$\frac{dV_\mathrm{P}(z)}{dz} = -\frac{P_z(z)}{\kappa_0 \kappa(z)} \quad \longrightarrow \quad V_\mathrm{P}(z) = -\int_0^z d\tilde{z}\, \frac{P_z(\tilde{z})}{\kappa_0 \kappa(\tilde{z})}, \tag{4.11}$$

This integral can be further simplified since $P_z(\tilde{z})/(\kappa_0\kappa(\tilde{z}))$ is a constant value which depends only on the material, i.e. the integral can be split up into contributions below, inside and above the InGaN layer:

$$V_\mathrm{P}(z) = \begin{cases} -z\frac{P_z(GaN)}{\kappa_0\kappa(GaN)} + C_1 & \forall\ 0 \leq z < z_b, \\ -z\frac{P_z(InGaN)}{\kappa_0\kappa(InGaN)} + C_2 & \forall\ z_b \leq z \leq z_t, \\ -z\frac{P_z(GaN)}{\kappa_0\kappa(GaN)} + C_3 & \forall\ z_t < z < z_{max}, \end{cases} \tag{4.12}$$

where z_{max} is the super cell size along z direction. With $P_z(\tilde{z})/(\kappa_0\kappa(\tilde{z}))$ having opposite signs in InGaN and GaN, it is possible to choose the integration constants C_1, C_2 and C_3, according to periodic boundary conditions, such that they allow a continuous behaviour of $V_\mathrm{P}(z)$ at the bottom and top interfaces z_b and z_t of the InGaN layer and furthermore allow $V_\mathrm{P} = 0$ at $z = 0$ and $z = z_{max}$. From Eq. (4.12) it can be concluded that the maximum (minimum) value of the polarisation potential increases (decreases) linearly with

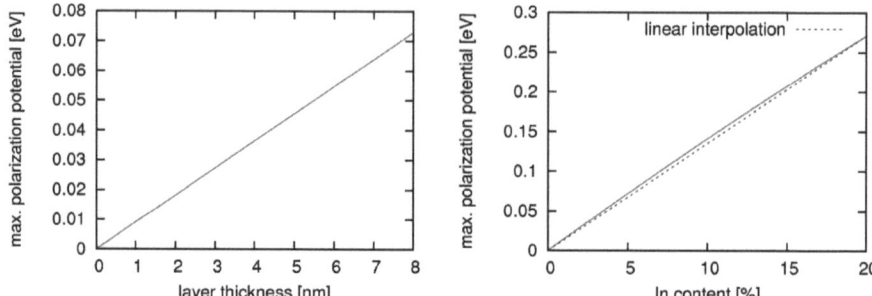

Figure 4.36: *The polarisation potential as a function of the In content in a 3.5 nm thick $In_xGa_{1-x}N$ layer (right) and as a function of the layer thickness for an In concentration of 10% (left). While the polarisation potential is a linear function of the layer thickness, only slight bowing is observed in the dependence of the In concentration.*

the layer thickness $d = z_t - z_b$ and almost linearly with the In concentration due to the weak parabolic bowing of ϵ_{zz} and the spontaneous polarisation P_{spont}. This can also be seen in the corresponding plots of the polarisation potential in Fig. 4.36. This result implies that the electron-hole overlap in quantum wells will increase for smaller thicknesses not only due to the stronger confinement but also due to the weaker effects of polarisation potentials.

The obtained strains and polarisation potentials can now be used within an eight band $\mathbf{k} \cdot \mathbf{p}$ model to perform systematic studies of the electronic properties of such quantum well systems.

InGaN layer thickness

The thickness of the InGaN layer in a polar InGaN/GaN superlattice influences both the confinement and thus the binding energies of the electrons and holes and the strength of the polarisation potential. The latter induces a charge carrier separation and a redshift of the emission wavelengths. To quantify these effects, studies of $In_{0.1}Ga_{0.9}N$ quantum wells embedded in a GaN matrix have been performed for thicknesses ranging from 0.25 nm to 8 nm. In order to provide data that are comparable with experimental observations, emission wavelengths have been derived from the electron and hole ground state binding energies ε_e and ε_h using the de Broglie relation:

$$\lambda = \frac{h}{\varepsilon_e - \varepsilon_h} \cdot c, \qquad (4.13)$$

where $h = 6.626 \cdot 10^{-34}$ Js is the Planck constant and $c = 2.998 \cdot 10^8$ m/s is the vacuum speed of light.

The emission wavelength resulting from the ground state electrons and holes and the corresponding charge carrier overlap is plotted in Fig. 4.37 as a function of the $In_{0.1}Ga_{0.9}N$ layer thickness. For an analysis of the influence of strain, piezoelectric potential and bulk electronic properties, the wavelength as well as the charge carrier overlap have been computed for different assumptions: Calculations for pure bulk electronic properties without contributions from strain and polarisation are the green curves in Fig. 4.37. The blue curve represents calculations performed using electronic and strain contributions, whereas the magenta curve has been calculated using only the electronic and the polarisation contributions. All three contributions are taken into account in the red curves in Fig. 4.37.

It can be seen that the emission wavelength converges for a large layer thickness, if no polarisation effects are taken into account (blue and green curves). For the calculation including only the electronic contributions, the wavelengths converge towards the wavelength that corresponds to the bulk $In_{0.1}Ga_{0.9}N$ band gap. If polarisation contributions are included in the calculation (red and magenta curves), the wavelength shows a linear increase with layer thicknesses above a thickness of approx. 4 nm. This linear behaviour results from the polarisation potential which scales linearly with the layer thickness.

The electron-hole overlap is found to decrease for larger layer thicknesses, resulting from the weaker localisation (see right plot in Fig. 4.37). Polarisation potentials increase the spatial separation of charge carriers in particular for large layer thicknesses. This is explained by the fact that with a larger thickness *i*) the spatial confinement decreases due to the spatial widening of the InGaN/GaN band offset potential and *ii*) the polarisation potential is growing linearly with the layer thickness. Correspondingly, the charge carrier overlap is only weakly affected by strain and polarisation effects for thin InGaN layers (below 3 nm).

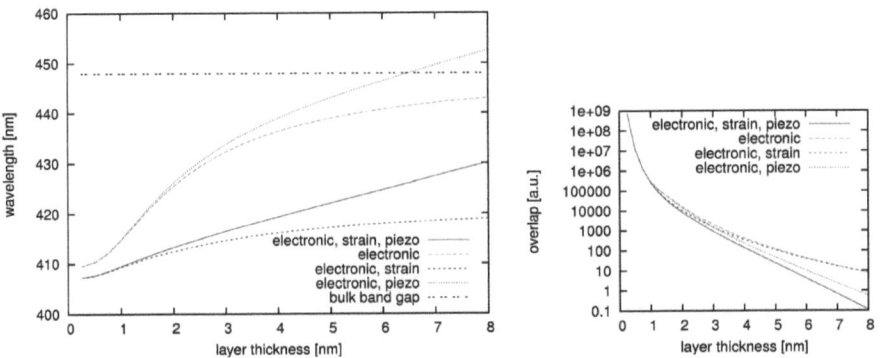

Figure 4.37: *Left: Emission wavelength derived from electron and hole ground states as a function of layer thickness. Right: Electron-hole overlap as a function of layer thickness.*

Influence of the In content in InGaN layers

Due to the large lattice mismatch between InN and GaN of around 10%, epitaxial growth of pure InN on a GaN surface becomes problematic. For typical light emission devices, ternary $In_xGa_{1-x}N$ alloys are used instead of pure InN. This enables not only an epitaxial growth due to the reduced lattice mismatch between GaN and an $In_xGa_{1-x}N$ layer, but furthermore allows to control the band gap of the $In_xGa_{1-x}N$ layer by varying the In content x. In practice, In contents in $In_xGa_{1-x}N$ up to 30% have been achieved [158].

Almost all parameters in the **k · p** and the continuum elasticity model for the ternary $In_xGa_{1-x}N$ alloy have been linearly interpolated between the bulk InN and GaN parameters. Only the band gap and the spontaneous polarisation have been interpolated by a parabolic approximation, as given in Eq. (4.18).

The emission wavelength corresponding to the difference between the electron and the hole ground state binding energies is plotted together with the charge carrier overlap in Fig. 4.38 as a function of the In content. Again, the colors in the plots represent the pure electronic (green), electronic and strain (blue), electronic and polarisation (magenta) and the model including all three contributions (red).

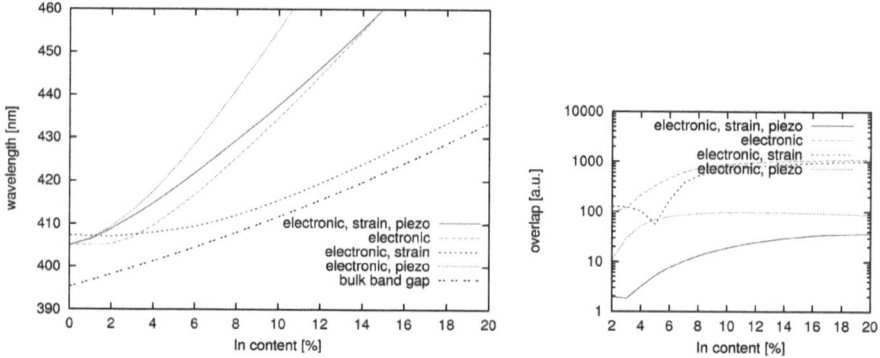

Figure 4.38: *Left: Emission wavelength derived from electron and hole ground states as a function of In content. Right: Electron-hole overlap as a function of In content.*

Please note the kink in the overlap for the case considering electronic and strain contributions beginning at an In content of 4%. This kink is explained by the strain-induced modifications of the three highest valence bands that can be seen in the eight band Hamiltonian in Eq. (A.1). The diagonal elements $\hat{H}_{33}$ to $\hat{H}_{88}$ contain the valence band edge and strain-induced contributions as potential terms (labeled here as V_{33} to V_{88}). However, these terms are not identical since the three highest valence bands are modified by the crystal-field and spin-orbit splitting and the strain-induced potential differences in a different manner.

The modified valence band edges in Eq. (A.1) are therefore:

$$V_{33} = V_{88} = E_{\text{vb}} + \Delta_{\text{cr}} + \frac{1}{3}\Delta_{\text{so}} + (D_1 + D_3)\epsilon_{zz} + (D_2 + D_4)(\epsilon_{xx} + \epsilon_{yy}),$$

$$V_{55} = V_{66} = E_{\text{vb}} - \Delta_{\text{cr}} + \frac{1}{3}\Delta_{\text{so}} + (D_1 + D_3)\epsilon_{zz} + (D_2 + D_4)(\epsilon_{xx} + \epsilon_{yy}),$$

$$V_{44} = V_{77} = E_{\text{vb}} + D_1\epsilon_{zz} + D_2(\epsilon_{xx} + \epsilon_{yy}).$$

The modified valence band edges are shown in Fig. 4.39 (left) as a function of In content. It is clearly visible that the $V_{33/88}$ and $V_{55/66}$ bands are separated only by the crystal field splitting and show a strong increase with the In content. The $V_{44/77}$ band edge increases only slightly with the In content due to the fact that $D_1\epsilon_{zz}$ has almost the same value but opposite sign as $D_2(\epsilon_{xx} + \epsilon_{yy})$ inside the InGaN layer. Correspondingly, the absolute valence band edge in the InGaN layer is represented by the $V_{44/77}$ band edge below an In concentration of 3.6% and by $V_{55/66}$ above this value. With the valence band edge, different kinetic parts of the Hamiltonian dominate the electronic structure leading to a modified localisation of the charge carriers whereas no such behaviour is seen in the emission wavelength. In particular, with $V_{44/77}$ being the maximum valence band edge, the dominating parts of the wave function are the Ψ_4 and Ψ_7 components. The kinetic part of the Hamiltonian is given here as

$$\hat{H}_{44/77} - V_{44/77} = \tilde{A}_1\partial_z^2 + \tilde{A}_2(\partial_x^2 + \partial_y^2), \tag{4.14}$$

which corresponds to a high effective hole mass in the in-plane directions (x and y) and a small effective mass in growth (z) direction. When the maximum valence band edge is represented by the $V_{55/66}$ band, the Ψ_5 and Ψ_6 components of the wave function become important. Here, the kinetic contribution is given by

$$\hat{H}_{55/66} - V_{55/66} = (\tilde{A}_1 + \tilde{A}_3)\partial_z^2 + (\tilde{A}_2 + \tilde{A}_4)(\partial_x^2 + \partial_y^2), \tag{4.15}$$

where the effective hole mass along z-direction becomes much larger and leads to a better localisation of hole states in the polarisation-modified valence band potential which increases the spatial separation of charge carriers and reduces the electron-hole overlap in the kink seen in Fig. 4.37. The contributions from the Ψ_4 and Ψ_7 are shown together with the contribution from the Ψ_5 and Ψ_6 components of the hole wave function in Fig. 4.39 (right) as a function of the In content. These contributions are defined such that the charge conservation is fulfilled:

$$\sum_{\sigma=1}^{8} |\Psi_\sigma|^2 = 1. \tag{4.16}$$

It is clearly visible that the hole wave function is dominated by the Ψ_5 and Ψ_6 components above an In concentration of approx. 4%. The kink in the charge carrier overlap is therefore explained by the crossing of valence bands and the resulting change of the importance of single components of the hole wave function at a certain In concentration. For the layer thickness chosen in this study, the emerging polarisation potentials lead to a reduction of

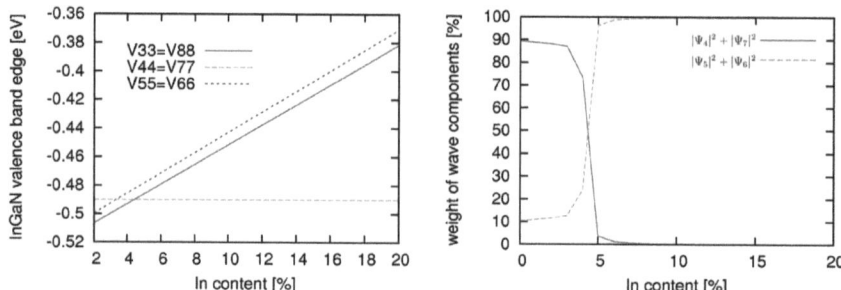

Figure 4.39: *Left: The three highest strain-modified valence band edges as a function of the In concentration. Right: Contributions from the Ψ_4 and Ψ_7 components (red) and the Ψ_5 and Ψ_6 components (green) to the hole wave function as a function of the In concentration.*

the electron-hole overlap of such strength that the reduced charge carrier overlap due to the above discussed effect becomes negligible. This can be seen in Fig. 4.38 (right), where no kink in the overlap is found for the electronic states computed including strain and polarisation potential. Due to the relationship between the layer thickness and the strength of the polarisation potential (see Eq. (4.12)), it can be expected that the reduced charge carrier overlap around an In concentration of 3.6% becomes important for smaller layer thicknesses, where the influence of polarisation potentials decreases. To confirm this assumption, the above calculations including strain and polarisation potential have been repeated for thinner InGaN quantum wells. The corresponding overlaps can be seen as a function of the In

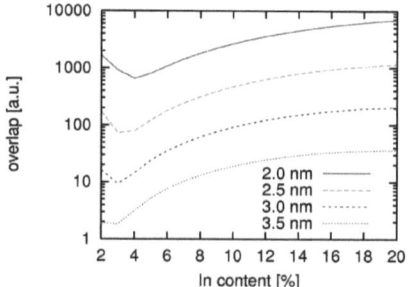

Figure 4.40: *Electron-hole overlap as a function of the In content in an InGaN/GaN quantum well for four different layer thicknesses.*

content in Fig. 4.40. It can be clearly seen that the reduction of the charge carrier overlap for an In content of about 4% is more pronounced in thin quantum wells, due to the reduced charge-separating influence of the polarisation potential. It can therefore be concluded, that this charge carrier separation becomes meaningful in particular in thin quantum wells of less than 2 nm thickness.

Summary: Polar InGaN/GaN quantum wells

Systematic studies have been performed concerning the electronic properties of polar InGaN quantum wells. Strain and polarisation potentials which can be calculated analytically for this one-dimensional problem have been used as an input for an eight band $\mathbf{k} \cdot \mathbf{p}$ model, leading to modifications of band edges and thus influencing the electronic properties.

Increasing the InGaN **layer thickness** for an In concentration of 10% from 0.25 nm to 8 nm results in an increasing emission wavelength, i.e. to a redshift due to the piezoelectric potential which grows linearly with the layer thickness. Simultaneously, the overlap between electrons and holes decreases due to the weaker localisation of charge carriers in a thicker layer and, moreover, due to the polarisation potential which spatially separates electrons and holes on the opposite interfaces of the InGaN layer.

The **In concentration** in $In_xGa_{1-x}N$ has been varied from $x = 0$ to $x = 0.2$, which covers the range of experimentally realisable concentrations. The wavelengths increase with higher In content, again leading to a redshift. This behaviour can be predicted mainly from the bulk $In_xGa_{1-x}N$ band gap. The overlap between electrons and holes converges for larger In concentrations. This is explained by the larger confinement of charge carriers due to an increasing conduction and valence band offset between the $In_xGa_{1-x}N$ layer and the GaN matrix on the one hand and the spatial separation of electrons and holes due to an increasing polarisation potential on the other hand.

The charge carrier overlap shows a kink for In concentrations of around 4%. This kink is in particular visible when switching off the polarisation potential in the electronic structure calculation. This kink was found to be the result of a crossing of the valence bands due to strain which leads to a change of the hole effective masses that dominate the hole state. The hole state localisation is therefore increased for an In concentration of 3.6% leading to a reduced electron-hole overlap which increases above this value due to stronger localisation resulting from the band offset between layer and matrix. This effect vanishes in thick quantum wells, where strong polarisation potentials induce a much stronger charge carrier separation. On the other hand, the reduced electron-hole overlap in InGaN layers of approx. 4% In content becomes important for quantum wells with less than 2 nm thickness, where the influence of polarisation potentials is comparatively small. In such systems, a reduced efficiency of light emission processes can be expected, due to the stronger influence of the strain-modified valence band edges on the spatial separation of electrons and holes.

4.3.2 Thickness fluctuations in nonpolar grown $In_{0.2}Ga_{0.8}N$ quantum wells

III-nitride semiconductors can be grown in the thermodynamically stable wurtzite phase or in a metastable zincblende phase [136]. The growth of wurtzite crystal structures in a polar direction is a well established procedure, but surfaces partially oriented in [0001]-direction will produce a polarisation potential resulting from the spontaneous polarisation [24, 25, 28]. The resulting electrostatic built-in field is estimated to be in the order of MV/cm in III-nitride systems [216, 248]. This results in a spatial separation of electrons and holes due to huge band bending effects and thus to a reduction of oscillator strength and radiative lifetimes [11, 12, 216] as well as to a redshift of the emission wavelength [92], as is studied in detail in the previous section.

In order to overcome this effect, much research effort has been invested in nonpolar growth processes of bulk GaN [98], GaN/AlN quantum wells [171], InGaN/GaN [48, 226] and InN/GaN [145] quantum wells where no [0001]-oriented surfaces occur as well as in comparative studies of polar, semipolar and nonpolar III-nitride quantum well structures [82, 126].

In the ideal case this procedure should eliminate electrostatic built-in fields resulting from spontaneous polarisation [4, 58, 241]. However, occuring thickness fluctuations in realistic nonpolar grown III-nitride quantum wells as reported, e.g., in Ref. [47] will again introduce [0001]-oriented surfaces resulting in a polarisation potential which again leads to a redshift of the emission wavelength and a spatial separation of charge carriers which reduces the emission efficiency.

While computationally expensive atomistic calculations can be used to investigate the effect of a few specific fluctuations on the electro-optical properties in III-nitride quantum well systems, a combination of the second-order continuum elasticity theory and the 8-band $\mathbf{k} \cdot \mathbf{p}$-formalism allows a systematic study of possible thickness and composition fluctuations with respect to their depth and width. Furthermore, this approach allows a systematic study of the influence of the shape and material composition of such a fluctuation. However, these studies are not subject of the present work.

The model system

In order to study the influence of thickness fluctuations on the built-in electrostatic potential and the charge carrier localisation in InGaN quantum wells, an 8 nm thick $In_{0.2}Ga_{0.8}N$ quantum well in a GaN matrix has been chosen, as observed e.g. by Kim and co-workers [121]. The growth direction is $x = [11\bar{2}0]$.

Due to the unavailability of detailed experimental studies concerning shape and characteristics of layer thickness fluctuations, this work restricts itself to a qualitative understanding of the effect of such structures. Therefore, the thickness fluctuation is modeled as a channel oriented along $y = [1\bar{1}00]$ with the width w in the $z = [0001]$ direction. The depth of the fluctuation, d, is varied between 0.25 nm and 8 nm. The model quantum well with fluctuation is shown in Fig. 4.41. Experimental observations so far found thickness fluctuations with a

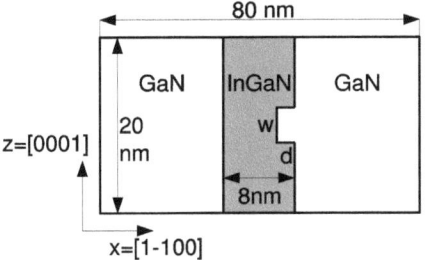

Figure 4.41: *Nonpolar quantum well with thickness fluctuation. The axes are labeled as:* $x = [11\bar{2}0]$, $y = [1\bar{1}00]$ *and* $z = [0001]$. *The cell size is 80 nm in growth-direction (x) and 20 nm in z-direction.*

depth of one lattice constant in polar quantum wells [47]. The calculations of strain, built-in potential and electronic properties were done using a mesh discretisation of 640 mesh points in x- and 160 mesh points in z-direction, which allows a sufficiently accurate resolution of the fluctuation in a quantum well of 8 nm thickness. The material parameters for InN and GaN are shown in Tab. 4.10. For the ternary alloy $In_xGa_{1-x}N$, most parameters are linearly interpolated for the In content x:

$$p(In_xGa_{1-x}N) = x \cdot p(InN) + (1-x) \cdot p(GaN), \quad (4.17)$$

where $p(In_xGa_{1-x}N)$ is the parameter to be interpolated. The band gap E_g and the spontaneous polarisation P_{sp}, require a quadratic interpolation in ternary alloys [27, 156]:

$$p(In_xGa_{1-x}N) = x \cdot p(InN) + (1-x) \cdot p(GaN) - x \cdot (1-x) \cdot b_p(InGaN). \quad (4.18)$$

Here, $b_p(In_xGa_{1-x}N)$ is the bowing of the parameter p.

Strain fields and polarisation potentials

Strain fields have been computed using second-order elasticity theory (see Chapter 2.3.4). Due to the epitaxial growth, the InGaN layer is strained to the bulk GaN lattice constants along y- and z-direction. For an unperturbed quantum well, this allows an analytic expression of the ϵ_{yy}, ϵ_{zz} and ϵ_{xx} strain components. Due to the lack of shear effects, non-diagonal strain components ϵ_{ij} with $i \neq j$ are zero. According to experimental observations, the GaN matrix is strain free, i.e. its lattice constants are the bulk lattice constants in all directions. The in-plane strains in the InGaN quantum well can thus be calculated solely from the lattice constants:

$$\epsilon_{yy} = \epsilon^0_{yy} = a(GaN) - a(In_{0.2}Ga_{0.8}N)/a(In_{0.2}Ga_{0.8}N) = -0.0218, \quad (4.19)$$

$$\epsilon_{zz} = \epsilon^0_{zz} = c(GaN) - c(In_{0.2}Ga_{0.8}N)/c(In_{0.2}Ga_{0.8}N) = -0.0196. \quad (4.20)$$

These strains are well below the critical values of the continuum elasticity theory, which does not apply for pure InN grown on GaN where these strains are in the order of 0.10 to 0.11.

Table 4.10: *Material parameters used in this study. Lattice and elastic constants as well as the values for spontaneous polarisation, piezoelectric tensor coefficients and dielectric constants were taken from Ref. [239]. The* $\mathbf{k} \cdot \mathbf{p}$ *parameters for effective masses, A_i's and band energies are taken from Ref. [202].*

Parameter	InN	GaN	Parameter	InN	GaN
a [Å]	3.545	3.189	E_g [eV]	0.69	3.24
c [Å]	5.703	5.185	E_{vb} [eV]	0.00	-0.50
C_{11} [GPa]	223	390	Δ_{cr} [eV]	0.040	0.010
C_{12} [GPa]	115	145	Δ_{so} [eV]	0.005	0.017
C_{13} [GPa]	92	106	$m_e^{\parallel}$ [m_0]	0.065	0.186
C_{33} [GPa]	224	389	$m_e^{\perp}$ [m_0]	0.068	0.209
C_{44} [GPa]	48	105	A_1	-15.803	-5.947
e_{15} [C/m^2]	0.264	0.326	A_2	-0.497	-0.528
e_{31} [C/m^2]	-0.484	-0.527	A_3	15.251	5.414
e_{33} [C/m^2]	1.060	0.895	A_4	-7.151	-2.512
P_{sp} [C/m^2]	-0.042	-0.034	A_5	-7.060	-2.510
ε_r	13.8	9.8	A_6	-10.078	-3.202

With the completely relaxed GaN matrix with $\epsilon_{xx} = \epsilon_{yy} = \epsilon_{zz} = 0$, the elastic energy in Eq. (2.43) is purely a function of the strain component ϵ_{xx} in growth direction in InGaN.

$$F(\epsilon_{xx}) = \frac{1}{2}L_x^0 L_y^0 L_z^0 \left[C_{11}(\epsilon_{xx}^2 + \epsilon_{yy}^0) + 2C_{12}\epsilon_{xx}\epsilon_{yy}^0 + 2C_{13}\epsilon_{zz}^0(\epsilon_{xx} + \epsilon_{yy}^0) + C_{33}(\epsilon_{zz}^0)^2 \right], \quad (4.21)$$

where the L_i^0's are the InGaN well dimensions along x-, y- and z-direction. The derivative $dF/d\epsilon_{xx}$ now yields the minimum elastic energy for the strain component ϵ_{xx} as:

$$\epsilon_{xx} = -\frac{C_{12}}{C_{11}}\epsilon_{yy}^0 - \frac{C_{13}}{C_{11}}\epsilon_{zz}^0 = 0.0140, \quad (4.22)$$

which reflects the Poisson ratio.

For the ideal quantum well, these strains can be easily derived analytically. A thickness fluctuation, however, induces rather complicated strain fields. For this purpose, the continuum elasticity model described in Sec. 2.3.4 was employed. It is essential to provide the ϵ_{yy}^0 and ϵ_{zz}^0 strain components for the elastic energy in Eq. (2.43). In principle, the continuum elasticity model applied should be able to compute the strain fields ϵ_{yy} and ϵ_{zz} including the ϵ_{yy}^0, ϵ_{zz}^0 components without providing these contributions explicitly. In the present case these strain components are constant inside the quantum well. Correspondingly, the displacements u_y and u_z which are related to the strains by

$$\epsilon_{yy} = \frac{\partial u_y}{\partial y} \quad \text{and} \quad \epsilon_{zz} = \frac{\partial u_z}{\partial z} \quad (4.23)$$

are linear functions. In a plane-wave picture, the displacements have to be periodic functions, which is not the case for linear functions. Therefore, the constant non-zero strains inside the quantum well have to be captured by the ϵ_{ij}^0 tensor.

Strain distributions for quantum wells with an unperturbed interface as well as for thickness fluctuations are shown in Fig. 4.42. Here, the diagonal strain components ϵ_{xx}, ϵ_{yy} and ϵ_{zz} are given for fluctuations of 3 nm width and depths of 0.25, 0.5, 1, 2 and 8 nm. For the shallow fluctuations (top right, middle left and middle right in Fig. 4.42), the deviation from the analytical solution for an unfluctuated quantum well is only a small increase of the absolute strain values in ϵ_{xx} and ϵ_{zz}. Above a fluctuation depth of 2 nm (Fig. 4.42, bottom left), a second peak of the ϵ_{xx} and ϵ_{zz} strains is observed. This peak is the result of a non-negligible strain inside the GaN filling the fluctuation, from the InGaN well at the [0001] and the [000$\bar{1}$] interface of the fluctuation. For a completely interrupted well (Fig. 4.42, bottom right), this contribution leads to a large strain inside the GaN between the InGaN [0001] and [000$\bar{1}$] interfaces.

Strain and spontaneous polarisation give rise to a built-in polarisation potential (see Eqs. (2.47) and (2.49)). The built-in polarisation potentials for a 3 nm wide fluctuation with depths of 0.25, 0.5, 1 and 2 nm depth are given in Fig. 4.43. The cell size of 20 nm along the [0001] direction is sufficiently large to prevent significant errors from periodic images.

It can be seen that a thickness fluctuation induces a non-negligible polarisation potential which modifies the conduction and valence band edges. Similar calculations have been made for 1 nm and 2 nm wide fluctuations. Additionally, the fluctuation's depth was systematically increased from 0.25 nm to 8 nm which corresponds to a complete interruption of the quantum well. The corresponding results are discussed in the following.

Spatial separation and binding energies of electrons and holes

The formation of polarisation potentials at the [0001]-oriented facets of the thickness fluctuation leads to a spatial separation of electron and hole states. The charge carrier localisation was computed using the $\mathbf{k} \cdot \mathbf{p}$ formalism. The electron (red) and hole (blue) ground state localisation is shown in Fig. 4.44 (left) for the 3 nm wide fluctuation for various depths. The polarisation potential obviously leads to a strong spatial separation of the charge carriers. The electron-hole overlap (Fig. 4.44 right) is computed via Eq. (2.53) for the electron and hole ground state as a function of the fluctuation depth d for $w=$1, 2 and 3 nm wide fluctuations. A significant reduction of the electron-hole overlap is found for thickness fluctuations of more than 1 nm depth. Single monolayer fluctuations do not introduce a considerable spatial separation of electrons and holes.

The binding energies of the electron and hole ground state are given for the different widths as a function of the fluctuation depth in Fig. 4.45. It is again observed that fluctuations of less than 1 nm have no significant influence on the binding energies. For larger fluctuations, the difference between electron and hole ground state is reduced, leading to larger wavelengths and thus to a redshift in light emission processes.

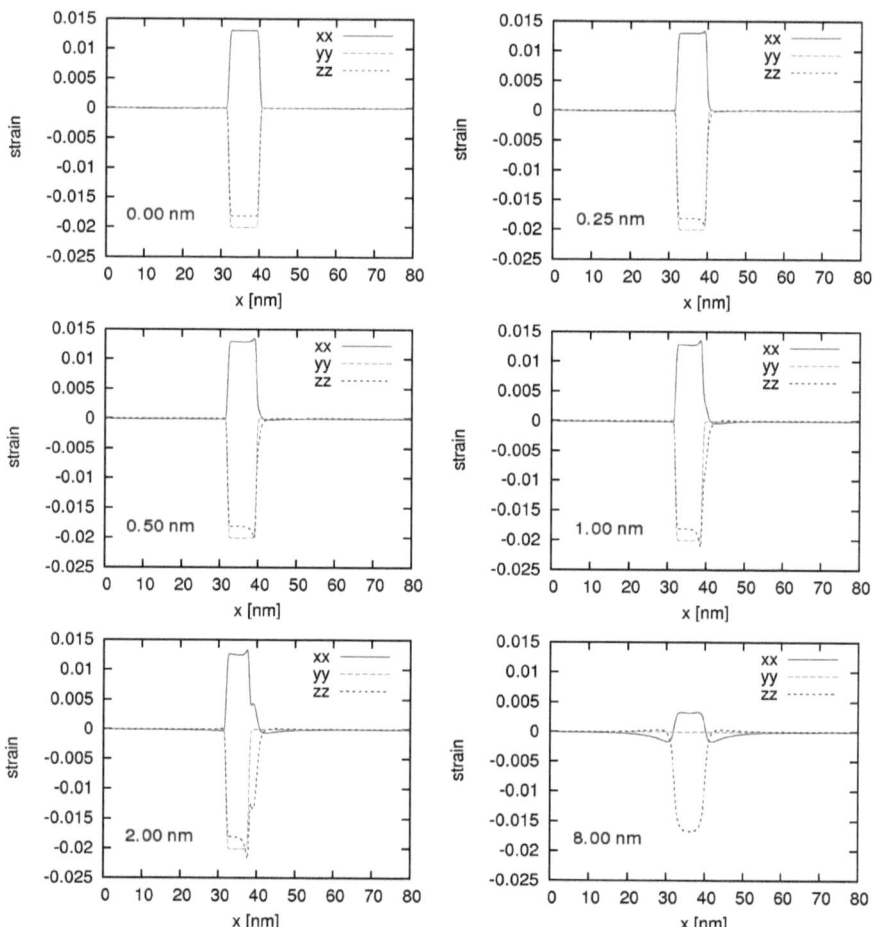

Figure 4.42: *Diagonal strain components in a nonpolar InGaN quantum well with thickness fluctuation of 3 nm width. The strains for the unfluctuated well are shown in the top left picture. The following pictures show the strains for fluctuation depths of: 0.25 nm (top right), 0.5 nm (middle left), 1 nm (middle right), 2 nm (bottom left) and 8 nm (bottom right).*

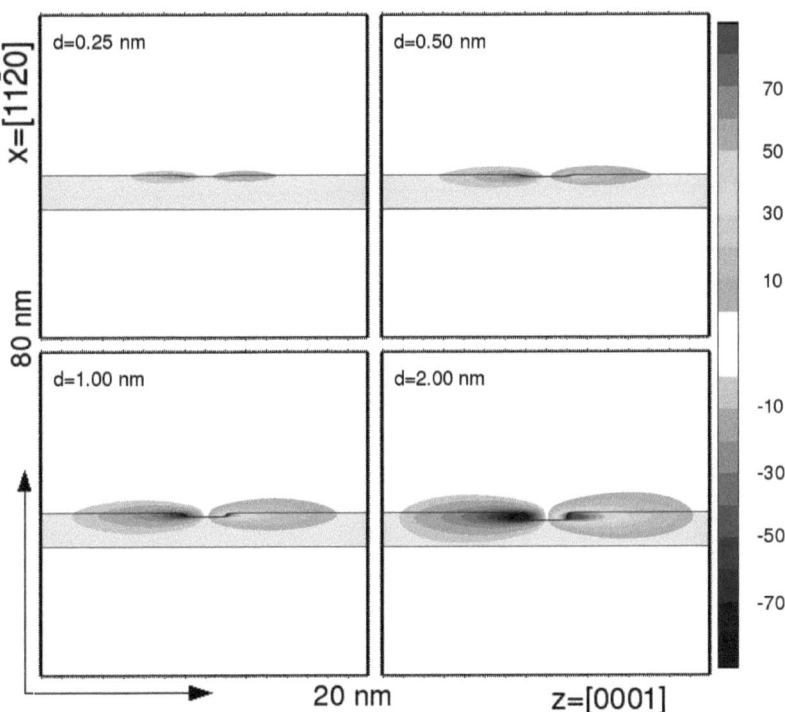

Figure 4.43: *Built-in polarisation potential in a nonpolar InGaN quantum well with thickness fluctuation of 3 nm width. The fluctuation depth is: 0.25 nm (top left), 0.5 nm (top right), 1 nm (bottom left) and 2 nm (bottom right).*

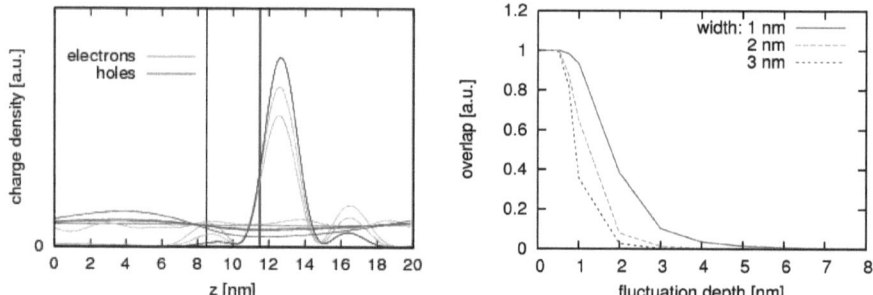

Figure 4.44: *Left: Charge carrier localisation for fluctuation depths of 0.25, 0.50, 1.00 and 2.00 nm in a 3 nm wide fluctuation. Right: electron-hole ground state overlap as a function of the fluctuation depth for 1, 2 and 3 nm wide fluctuations.*

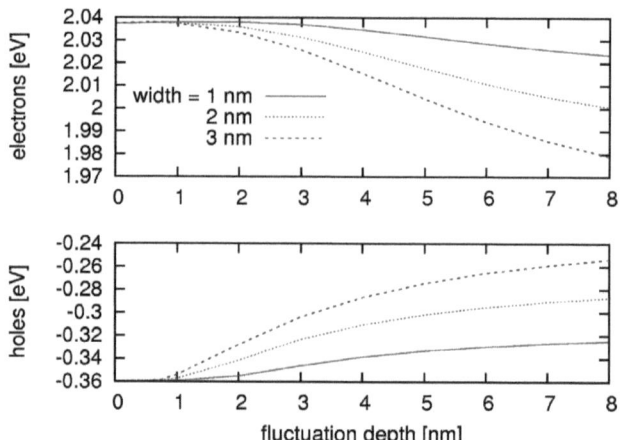

Figure 4.45: *Binding energies of the electron (top) and hole (bottom) ground state as a function of the fluctuation depth for 1, 2 and 3 nm wide fluctuations.*

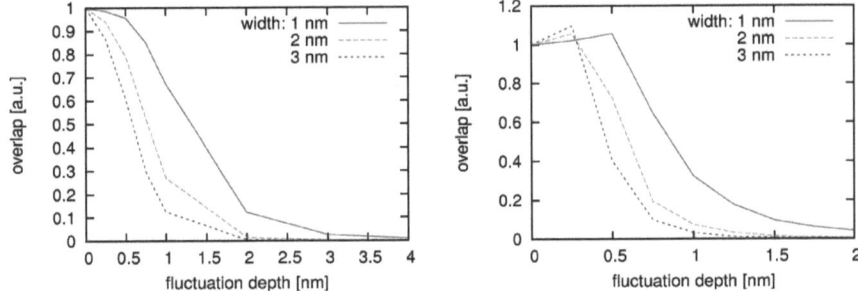

Figure 4.46: *Charge carrier overlap in a 4 nm (left) and a 2 nm (right) thick quantum well as a function of the fluctuation depth d for different widths.*

Influence of the layer thickness

The previous calculations have been repeated for InGaN quantum wells of 4 nm and 2 nm thickness to investigate the influence of thickness fluctuations in thinner quantum wells, as, e.g., observed in Ref. [47]. The strain fields and polarisation potentials do not show a qualitatively different behaviour for thinner quantum wells in comparison to those in the above studied system of 8 nm thickness. Nevertheless, the reduced layer thickness is expected to influence the charge carrier overlap as well as the binding energies of the confined electron and hole states. Fig. 4.46 shows the charge carrier overlap in a 4 nm and a 2 nm thick quantum well with a fluctuation width of 1, 2 and 3 nm. The binding energies of electrons and holes are given in Figs. 4.47 and 4.48, respectively. It can be seen, that the electron-hole overlap decreases slower for smaller layer thicknesses. This can be explained by the stronger overlap in thinner quantum wells which is purely a result of the reduced dimensions. Moreover, it can be seen in the 2 nm thick quantum well, that the charge carrier overlap increases, in particular for narrow fluctuations, for a depth of 0.25 nm before it decreases for deeper fluctuations. This is also a result of the strong confinement together with the polarisation-induced localisation of electrons and holes on opposite sides of the fluctuation with only a small distance in between (1 to 3 nm). This effect is not observed in quantum wells of larger thickness. Furthermore, shallow fluctuations, such as experimentally observed of ≈0.5 nm depth, do not have a significantly larger influence in thin quantum wells, as can be seen when comparing the overlaps for the different well thicknesses. For all three layer thicknesses considered in this work, the overlap is given as a function of the relative fluctuation depth with respect to the layer thickness in Fig. 4.49. It is observed, that the charge carrier overlap is not similar for the same relative fluctuation depth and thus shows that the influence of shallow fluctuations is not increasing significantly for smaller layer thicknesses. The binding energies behave similar in the studied quantum wells with different thicknesses. The energy difference between the electron and the hole ground state

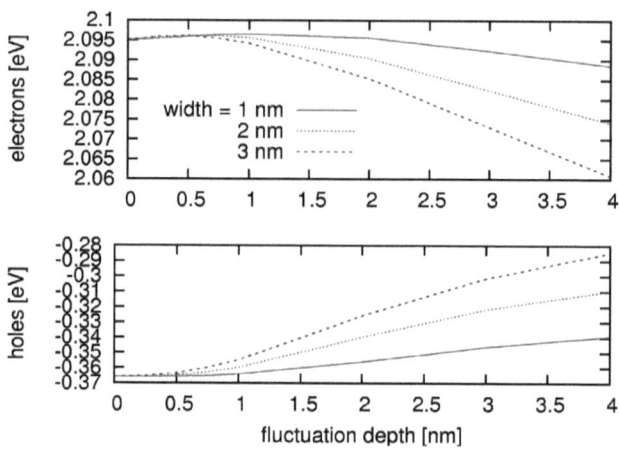

Figure 4.47: *Binding energies of the electron (top) and hole (bottom) ground state as a function of the fluctuation in a 4 nm thick quantum well.*

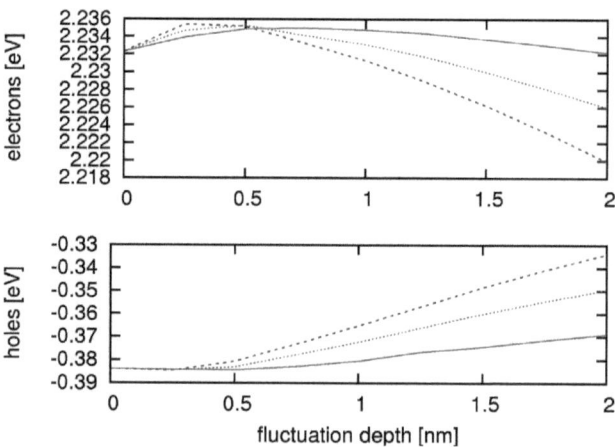

Figure 4.48: *Binding energies of the electron (top) and hole (bottom) ground state as a function of the fluctuation in a 2 nm thick quantum well.*

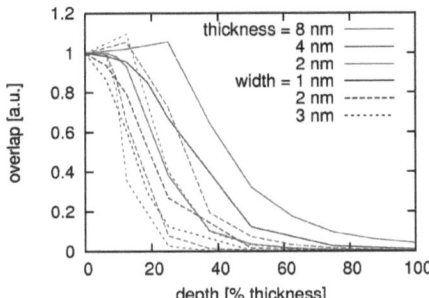

Figure 4.49: *Charge carrier overlap in an 8 (red), 4 (blue) and 2 nm (magenta) thick quantum quantum well of 1 (solid), 2 (dashed) and 3 nm width (dotted lines) as a function of the relative fluctuation depth with respect to the total quantum well thickness.*

is found to increase for smaller well thicknesses, as expected due to the stronger confinement. Furthermore, it is observed that the ground state energy of the electron increases for shallow fluctuations before it decreases due to the polarisation potential leading to the well known red shift in the emission wavelength. This increase of the electron binding energy is more pronounced in thin quantum wells and can be attributed to the stronger confinement due to the thickness fluctuation. A similar behaviour is observed for the hole ground state. However, this effect is much weaker.

Conclusions: Nonpolar InGaN quantum wells

The influence of thickness fluctuations in nonpolar grown $In_{0.2}Ga_{0.8}N$ quantum wells on the electronic properties was investigated. This study provides a qualitative understanding of polarisation-induced charge carrier separation due such thickness fluctuations. Within the studied quantum well systems of 2, 4 and 8 nm thickness, significant reductions of the electron-hole overlap and changes of the corresponding binding energies occur only for fluctuation depths of more than 1 nm, whereas experimental observations so far found maximum depths of two monolayers (0.52 nm) [47]. In particular, it was found that even in thin quantum wells the effect of realistic thickness fluctuations is not large enough to induce a dramatic reduction of the electron-hole overlap. It can therefore be concluded that the experimentally confirmed thickness fluctuations do not induce a major spatial separation of electrons and holes for InGaN layers of 8 nm thickness.

There is much evidence from experiments, that charge carriers localise due to alloy composition fluctuations arising from low indium solubility in GaN in nonpolar grown InGaN quantum wells [49, 138, 142, 168, 178], which is a well-known feature also in polar grown quantum wells [46, 105, 175] and other III-V superlattice systems [253]. However, the influence of composition fluctuations is not subject of the present study. Within this work, only single-particle states have been investigated. For practical applications, energetically lower lying states can possibly fill up the existing polarisation potentials and therefore neutralise the existing polarisation potentials leading to a smaller or completely vanishing spatial separation of electrons and holes.

Chapter 5
Summary and outlook

The aim of this work was the investigation of a wide range of III-nitride nanostructures that are promising candidates for the development or improvement of future light emitting devices. A theoretical study of a large number of nanostructured systems and possible modifications requires both, a model which provides an accurate and straightforward description of these systems, and an efficient implementation of this model to achieve the correspondingly large throughput of calculations with reasonable computational costs.

For this purpose, the well established $\mathbf{k} \cdot \mathbf{p}$ formalism as well as a second-order continuum elasticity model have been formulated and implemented in a novel way using a plane-wave based formulation. We were able to demonstrate that a plane wave-formulation of these two methods, that are dominated by differential equations, allows a computationally much cheaper formulation of the required gradient operators, than the traditional implementation using a finite-differences picture in real space. It has been shown that besides the efficient formulation of gradient operators in reciprocal space, the formulation of a Poisson solver can be done straightforwardly in a plane-wave picture and the accuracy of the performed calculations can be controlled directly via the plane-wave cutoff energy, i.e., the number of plane waves employed for the description of the system. Moreover, the implementation of a plane-wave formulated continuum elasticity and $\mathbf{k} \cdot \mathbf{p}$ model in the existing plane-wave software library S/PHI/nX allows to employ existing, highly optimised minimisation techniques such as the preconditioned conjugate-gradient scheme. For this purpose, these minimisation algorithms including their corresponding preconditioner schemes have been modified to fit the specific requirements of the continuum-elasticity model and the eight band $\mathbf{k} \cdot \mathbf{p}$ formalism.

The implementation of the eight band $\mathbf{k} \cdot \mathbf{p}$ model employed in this work is easily extendable to a higher number of involved bands. However, a **comparison with atomistic models**, i.e., an empirical tight-binding approach and its effective bond orbital model simplification, has shown that taking only the lowest conduction band and the three highest valence bands into account allows to compute electronic properties with high accuracy even for nanostructures with characteristic dimensions of only a few nanometers.

The elastic properties of semiconductor nanostructures were obtained in this work from a second-order continuum elasticity model. In order to verify the validity of an approach that is restricted to second-order elastic effects, a comparison has been performed to a **third-order elasticity** model. The differences in strain and polarisation fields and, moreover, in the resulting electronic properties are extremely small which, in conclusion, allows to restrict the elasticity model employed within this work to second-order terms. In particular, larger effects of third-order elasticity might become meaningful in regions where much larger strains (i.e. above 10%) occur. For the systems studied within the present work using the second-order elasticity model, such large strains did not occur. Anyway, a continuum elasticity model is not the best choice for systems where such large strains are present.

The implemented continuum elasticity and $\mathbf{k}\cdot\mathbf{p}$ formalism allows to study a wide range of III-nitride semiconductor nanostructures, which are promising candidates for an application in light emission devices such as lasers and LEDs, in a highly efficient manner.

Systematic studies of **wurtzite GaN/AlN quantum dots on polar substrates** have been performed to provide general information about the influence of shape, size and material composition of quantum dots. One of the key results of this study is that the parameters, which allow significant modifications of the electron and hole binding energies and hence, the emission spectrum, are the size and the material composition. The influence of the quantum dot shape is found to be rather small. This independence on the shape is a fortunate situation, since experimental experience shows that the shape of quantum dots is rather difficult to control in an epitaxial growth process.

Within this work, special attention has been paid to the influence of polarisation effects in wurtzite III-nitride quantum well and quantum dot systems. **GaN/AlN quantum dots grown on nonpolar surfaces** have been experimentally observed, recently and are promising candidates for light emission devices with reduced spatial separation of electrons and holes which induces weak recombination rates in polar GaN/AlN quantum dots. Within this work the charge carrier separation due to polarisation potentials was found to be even larger than in polar quantum dots if one considers the experimentally observed system geometries and dimensions. Reducing the characteristic dimensions was found to lead to a dramatic increase of the charge carrier overlap and thus to improved recombination rates. This effect is much stronger than in quantum dots grown on polar surfaces, which allows the conclusion that the size of nonpolar quantum dots is the key parameter to improve the light emission efficiency.

The eight band $\mathbf{k} \cdot \mathbf{p}$ model has been applied to investigate the **charge carrier localisation around screw dislocations** due to shear strains in bulk GaN. The strain values have been derived analytically and were used as an input in the $\mathbf{k} \cdot \mathbf{p}$ Hamiltonian used to calculate the electronic structure of the system. It was found that the electron states are not visibly affected by the shear strains and thus show no localisation around the screw dislocation. For the hole states, however, a strong influence of the shear strains was observed, leading to a localisation of a large number of hole states around a shear dislocation. This will induce electron-hole recombination processes in the area of shear dislocations. These results confirm the interpretation of some experiments that find shear dislocations in bulk GaN to act as, commonly non-radiative, recombination centers.

GaN quantum wires in vacuum have been studied to predict the dependence of electronic properties of such structures on their diameter in order to identify promising quantum wire systems for novel light emitting devices. Being completely relaxed during the growth process, these quantum wires show no significant strains. Therefore, pure bulk properties dominate the electronic structure of these wires. Correspondingly, the electron and hole binding energies converge towards the conduction and the valence band edge for quantum wire diameters above a certain size. Within this work, a diameter of 10 nm was found to be sufficient to obtain pure bulk properties with only small quantisation effects. In order to introduce quantum effects due to charge carrier localisation, the diameter of GaN quantum wires has to be below this limit. Since the size of the smallest wires that were experimentally observed so far already have a diameter close to this value, this requirement is considered to be experimentally achievable.

A systematic study of **polar InGaN/GaN** quantum wells has been performed to investigate the influence of quantum well thickness and In content on the polarisation-induced spatial separation and the binding energies of electrons and holes. To allow a comparison to experimental data, light emission wavelengths have been estimated from the energy difference between the electron and the hole ground state. Increasing the well thickness was found to result in a decrease of the electron-hole overlap and an increase of the emission wavelength, resulting from the polarisation potential which increases linearly with the layer thickness. Increasing the In content x in $In_xGa_{1-x}N$ quantum wells embedded in a GaN matrix again increases the emission wavelength. The charge carrier localisation on the other hand, is found to converge, which is explained by the interplay between stronger localisation due to a higher band offset barrier for larger In concentrations and the increasing spatial separation due to the stronger polarisation potential. Furthermore, it was found that the charge carrier overlap is reduced by strain effects for an In concentration of 3.6%. This results from a crossing of the three highest valence bands at this value and the corresponding effective masses that dominate the Hamiltonian. However, polarisation contributions lead to a reduction of this effect in quantum wells above 3 nm thickness, such that it is only observed for the case that polarisation potentials are neglected while strain contributions are taken into account or for thin quantum wells below 3 nm thickness.

The spatial separation of electrons and holes observed in polar InGaN quantum wells due to polarisation potentials can be circumvented by using **nonpolar quantum wells** instead. Having no interfaces oriented along the polar [0001] direction, no polarisation fields and thus no spatial separation of electrons and holes should occur. In realistic nonpolar InGaN quantum wells, however, fluctuations of the In concentration and the layer thickness again induce polarisation effects. Within this work, the influence of layer thickness fluctuations on the charge carrier localisation was investigated. It was found that layer thickness fluctuations above 1 nm of depth in an 8 nm thick quantum well can in fact dramatically reduce the electron-hole overlap, i.e., such a thickness fluctuation induces polarisation potentials which lead to a localisation of electrons on the one and holes on the other side of the fluctuation. Furthermore, a redshift of the emission wavelength is found. A fluctuation of 1 nm depth, however, corresponds to 4 monolayers of InGaN, which is so far not reported as experimental observation. Previous experimental studies so far found that thickness fluctuations in InGaN quantum wells have a maximum depth of 2 monolayers, which is not expected to induce a significant charge carrier separation, according to the present work. Moreover, the weak effect of one or two monolayers deep thickness fluctuations does not significantly increase for thin quantum well structures.

In summary, it has been shown that a nonpolar growth direction is in fact able to reduce the spatial separation of electrons and holes in III-nitride nanostructures, thus allowing to increase the light emission efficiency of such devices. This work provides the theoretical background to understand the charge carrier separation as well as detailed information about the key parameters to improve this efficiency for a wide scale of investigated systems.

Within the present work electronic properties of various III-nitride nanostructures have been investigated. Electronic wave functions and binding energies can now serve as an input for many-particle calculations which allow to compute the optical properties of the systems investigated within this work.

The continuum elasticity model employed here was used to compute strain and polarisation potentials as a contribution to the calculation of the electronic structure. Of course, this model is not limited to semiconductor materials and can therefore be employed for other epitaxially grown crystalline materials, e.g. for metallic systems. Furthermore, this model has no upper limit of the cell dimensions and can therefore be employed likewise to macroscopic systems.

The applied $\mathbf{k} \cdot \mathbf{p}$ model employing the lowest conduction and the three highest valence bands each with separate spin up and spin down components requires an eight band basis and was found to provide an excellent description of the electronic structure of III-nitride systems. Within this model, the parameter set is fitted to the band structure in the vicinity of the Γ-point. Since in InGaN and GaN a direct band gap at the Γ point is found, an accurate description of bulk-like systems is commonly expected and was likewise found for nanostructured systems within this work. This good agreement, however, cannot be expected for every semiconductor material. Additional minima in the band structure, in particular in case of a semiconductor material with an indirect band gap, reduce the importance of the Γ-point and make an eight band description inaccurate. For such bulk materials as well as for nanostructures, it is possible to employ $\mathbf{k} \cdot \mathbf{p}$ models with a higher number of bands which provide a more accurate description of the band structure also in regions far off the Γ point [78, 200, 260]. The implementation of such a model can be done in a plane-wave formulation similar to the eight band model described within this work.

In this work, special attention was paid to the influence of the strong polarisation effects that occur in wurtzite III-nitride nanostructures. In particular, studies of nonpolar quantum wells and quantum dots have been performed. However, there is still a number of possible extensions to the model systems investigated in this work. Concerning nonpolar InGaN/GaN quantum wells, fluctuations of the In content are suspected to act as quantum-dot like recombination centers allowing a more efficient light emission [131, 169]. Therefore, varying In profiles in InGaN quantum wells, thus simulating In clustering and the formation of zero-dimensional localisation centers are an important step towards a better understanding of the optoelectronic properties of InGaN quantum well structures. Thickness fluctuations have been modeled as a two-dimensional problem and are therefore reduced to channel-like structures of homogeneous depth and width with an infinite length. Realistic structures are expected to be more complicated, i.e. the depth of such fluctuations might not be uniform and also the shape of such structures can be arbitrary. It is therefore important to gain more insight into realistic thickness fluctuations in nonpolar quantum wells and to investigate such systems based on experimental observations in systematic theoretical studies.

The implemented $\mathbf{k} \cdot \mathbf{p}$ formalism was shown to be in excellent agreement with atomistic ETBM calculations for GaN quantum dots with less than 2 nm height. Nevertheless, atomistic calculations are more reliable for small quantum dots due to the larger influence of single atomic effects which are not captured within a continuum approach. Additionally the computational effort of atomistic simulations is more reasonable for smaller structures. Employing parameters from highly accurate first principles G_0W_0 calculations, atomistic models, e.g. a tight-binding model where the required tight-binding parameters are constructed from maximally localised Wannier functions [154, 217, 240], can be combined consistently with continuum $\mathbf{k} \cdot \mathbf{p}$ models in a multiscale approach. The ETBM can then be used for studies of small nanostructures whereas the continuum $\mathbf{k} \cdot \mathbf{p}$ model is employed for larger structures. Such a combined tool allows a consistent modeling of III-nitride quantum dots from atomistic to mesoscopic length scales with a maximum of efficiency and accuracy.

Appendix A

The 8 × 8 k · p-Hamiltonian for wurtzite systems

Following Refs. [7] and [53], the eight band **k · p**-Hamiltonian is formulated for wurtzite systems in the basis set:

$$|\Phi\rangle = \begin{pmatrix} i|\Phi_s \uparrow\rangle \\ i|\Phi_s \downarrow\rangle \\ -\frac{1}{\sqrt{2}}|(\Phi_{p_x} + i\Phi_{p_y}) \uparrow\rangle \\ |\Phi_{p_z} \uparrow\rangle \\ \frac{1}{\sqrt{2}}|(\Phi_{p_x} - i\Phi_{p_y}) \uparrow\rangle \\ -\frac{1}{\sqrt{2}}|(\Phi_{p_x} + i\Phi_{p_y}) \downarrow\rangle \\ |\Phi_{p_z} \downarrow\rangle \\ \frac{1}{\sqrt{2}}|(\Phi_{p_x} - i\Phi_{p_y}) \downarrow\rangle \end{pmatrix}.$$

The Hamiltonian can be written as:

$$\hat{H}^{8\times 8} = \hat{H}^{8\times 8}_{\text{unstrained}} + \hat{H}^{8\times 8}_{\text{strain}} + V_{\text{P}},$$

with:

$$\hat{H}^{8\times 8}_{\text{unstrained}} = \begin{pmatrix} \hat{H}_c & \hat{H}_s \\ \hat{H}_s^\star & \hat{H}_v \end{pmatrix} = \begin{pmatrix} S & 0 & -V & U & V^\star & 0 & 0 & 0 \\ 0 & S & 0 & 0 & 0 & -V & U & V^\star \\ -V^\star & 0 & F & -H^\star & -K^\star & 0 & 0 & 0 \\ U & 0 & -H & \lambda & H^\star & \Delta & 0 & 0 \\ V & 0 & -K & H & G & 0 & \Delta & 0 \\ 0 & -V^\star & 0 & \Delta & 0 & G & -H^\star & -K^\star \\ 0 & U & 0 & 0 & \Delta & -H & \lambda & H^\star \\ 0 & V & 0 & 0 & 0 & -K & H & F \end{pmatrix}.$$

(A.1)

The 2 × 2 $\hat{H}_c$ denotes the conduction band, the 6 × 6 operator $\hat{H}_v$ the valence bands and $\hat{H}_s$ introduces the coupling between CB and VB.

$$S = E_{\text{cb}} + A_1' \mathbf{k}_z^2 + A_2' \left(\mathbf{k}_x^2 + \mathbf{k}_y^2\right),$$

$$
\begin{aligned}
F &= \Delta_1 + \Delta_2 + \lambda + \theta, \\
G &= \Delta_1 - \Delta_2 + \lambda + \theta, \\
\lambda &= \frac{\hbar^2}{2m_0}\left(\tilde{A}_1 \mathbf{k}_z^2 + \tilde{A}_2\left[\mathbf{k}_x^2 + \mathbf{k}_y^2\right]\right) + E_{\text{vb}}, \\
\theta &= \frac{\hbar^2}{2m_0}\left(\tilde{A}_3 \mathbf{k}_z^2 + \tilde{A}_4\left[\mathbf{k}_x^2 + \mathbf{k}_y^2\right]\right), \\
K &= \frac{\hbar^2}{2m_0}\tilde{A}_5\left(\mathbf{k}_x + i\mathbf{k}_y\right)^2, \\
H &= \frac{\hbar^2}{2m_0}\tilde{A}_6 \mathbf{k}_z\left(\mathbf{k}_x + i\mathbf{k}_y\right), \\
U &= i\mathbf{k}_z P_1, \\
V &= i\left(\mathbf{k}_x + i\mathbf{k}_y\right) P_2, \\
\Delta &= \sqrt{2}\Delta_3,
\end{aligned}
$$

with:

$$
\begin{aligned}
A'_1 &= \frac{\hbar^2}{2m_e^{\parallel}} - \frac{P_1^2}{E_g}, & A'_2 &= \frac{\hbar^2}{2m_e^{\perp}} - \frac{P_2^2}{E_g}, \\
\tilde{A}_1 &= A_1 + \frac{2m_0}{\hbar^2}\frac{P_2^2}{E_g}, & \tilde{A}_2 &= A_2, \\
\tilde{A}_3 &= A_3 - \frac{2m_0}{\hbar^2}\frac{P_2^2}{E_g}, & \tilde{A}_4 &= A_4 + \frac{m_0}{\hbar^2}\frac{P_1^2}{E_g}, \\
\tilde{A}_5 &= A_5 + \frac{m_0}{\hbar^2}\frac{P_1^2}{E_g}, & \tilde{A}_6 &= A_6 + \frac{\sqrt{2}m_0}{\hbar^2}\frac{P_1 P_2}{E_g}, \\
P_1^2 &= \frac{\hbar^2}{2m_0}\left(\frac{m_0}{m_e^{\perp}} - 1\right)\frac{(E_g + \Delta_1 + \Delta_2)(E_g + 2\Delta_2) - 2\Delta_3^2}{E_g + 2\Delta_2}, \\
P_2^2 &= \frac{\hbar^2}{2m_0}\left(\frac{m_0}{m_e^{\parallel}} - 1\right)\frac{E_g[(E_g + \Delta_1 + \Delta_2)(E_g + 2\Delta_2) - 2\Delta_3^2]}{(E_g + \Delta_1 + \Delta_2)(E_g + \Delta_2) - \Delta_3^2}, \\
\Delta_1 &= \Delta_{\text{cr}} \quad \text{and} \quad \Delta_2 = \Delta_3 = \frac{1}{3}\Delta_{\text{so}}.
\end{aligned}
$$

The strain induced contribution $\hat{H}_{\text{strain}}$ reads:

$$
\hat{H}_{\text{strain}} = \begin{pmatrix}
s & 0 & 0 & 0 & 0 & 0 & 0 & 0 \\
0 & s & 0 & 0 & 0 & 0 & 0 & 0 \\
0 & 0 & f & -h^\star & -k^\star & 0 & 0 & 0 \\
0 & 0 & -h & \lambda_\epsilon & h^\star & 0 & 0 & 0 \\
0 & 0 & -k & h & f & 0 & 0 & 0 \\
0 & 0 & 0 & 0 & 0 & f & -h^\star & -k^\star \\
0 & 0 & 0 & 0 & 0 & -h & \lambda_\epsilon & h^\star \\
0 & 0 & 0 & 0 & 0 & -k & h & f
\end{pmatrix} \quad (\text{A.2})
$$

where:
$$\begin{aligned}
s &= a_2(\epsilon_{xx} + \epsilon_{yy}) + a_1\epsilon_{zz}, \\
\lambda_\epsilon &= D_1\epsilon_{zz} + D_2(\epsilon_{xx} + \epsilon_{yy}), \\
\theta_\epsilon &= D_3\epsilon_{zz} + D_4(\epsilon_{xx} + \epsilon_{yy}), \\
f &= \lambda_\epsilon + \theta_\epsilon, \\
k &= D_5(\epsilon_{xx} + 2i\epsilon_{xy} - \epsilon_{yy}), \\
h &= D_6(\epsilon_{zx} + i\epsilon_{yz}).
\end{aligned} \quad (A.3)$$

The parameters $m_\perp$, $m_\parallel$, A_{1-6}, $E_g = E_{cb} - E_{vb}$, $\Delta_{1,2,3}$, a and D_{1-6} are again spatially dependent material properties.

Appendix B

Convergence tests and benchmarks

B.1 Mesh accuracy

For the zincblende GaN dot in Sec. 4.1.1, the influence of the mesh accuracy on the electron binding energies has been investigated. Fig. B.1 shows a plot of the four lowest electron binding energies as a function of mesh points along one direction. These calculations were performed in an effective mass approximation, where the conduction and the valence band part of the Hamiltonian in Eq. (2.34) are decoupled. It can be seen, that for the chosen system, 50 to 60 mesh points in each direction are sufficient to provide converged binding energies. For most quantum dot systems investigated in this work, a resolution of $80 \times 80 \times 80$ mesh points has been employed.

B.2 Mesh softening

To model the influence of interatomic diffusion on the electronic structure of a given nanostructure, a diffusion equation can be employed to soften the interfaces between the materials in the system. Furthermore, such a softening removes abrupt changes in the real-space material parameters, which are difficult to describe using plane waves. Therefore, a softening of the interfaces between the materials is expected to reduce the number of minimisation steps required to achieve a given accuracy in the calculation of electronic states.

A system which consists of two components is described as a function $0 \leq s(\mathbf{r}) \leq 1$, where 0 and 1 represent the pure materials. The diffusion of the interfaces is now achieved via

$$s^{j+1}(\mathbf{r}) = s(\mathbf{r})^j + \frac{1}{6\tau} \sum_{i=1}^{6} \Delta s_i^j, \qquad (B.1)$$

where $s_i^j = s^j(\mathbf{r}_i) - s^j(\mathbf{r})$ with $\mathbf{r}_i = \mathbf{r} \pm d\mathbf{x}, d\mathbf{y}, d\mathbf{z}$ is the difference between the material composition at $\mathbf{r}$ and its neighbouring meshpoints. $d\mathbf{x}$, $d\mathbf{y}$ and $d\mathbf{z}$ denote translations of one mesh point in x, y and z direction. The parameter τ is a diffusion parameter which decides how much intermixing between $s(\mathbf{r})$ and its neigbouring mesh points happens with

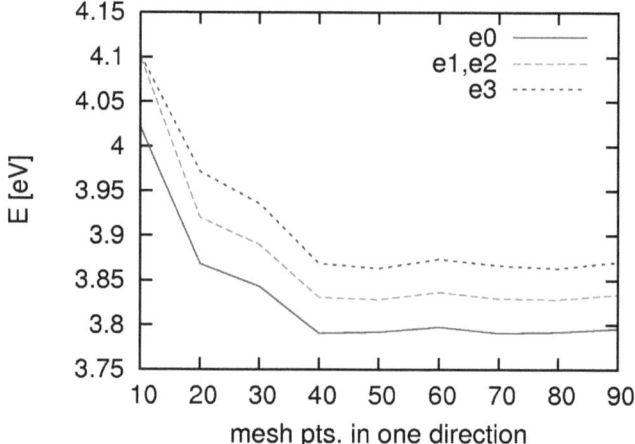

Figure B.1: *Binding energies of the four lowest electron states in a zincblende GaN quantum dot in AlN for different mesh accuracies $n \times n \times n$.*

each softening step and j is the number of softening steps. The softening of the mesh is done by looping over Eq. (B.1) for a given number of softening steps. In order to investigate the influence of mesh softening on the number of required convergence steps, a value of $\tau = 100$ has been chosen to ensure that the influence of the softening on the electronic states is minimal. The number of required convergence steps per state is then shown as a function of softening steps applied to the system. As a test case, a hexagonal based GaN quantum dot in AlN has been chosen.

It can be seen in Fig. B.2, that the influence of such a softening on the electronic structure is rather small, i.e. in the range of single meV. The number of minimisation steps to achieve energy convergence is reduced with employing only a small number of softening steps to 2/3 of the required minimisation steps. It is important to mention that small parameters for τ as well as large number of softening steps lead to strong modifications of the model system. Therefore, a softening of the interfaces in a model system should be applied only if the influence of τ and the number of softening steps on the electronic properties of the model system has been carefully evaluated.

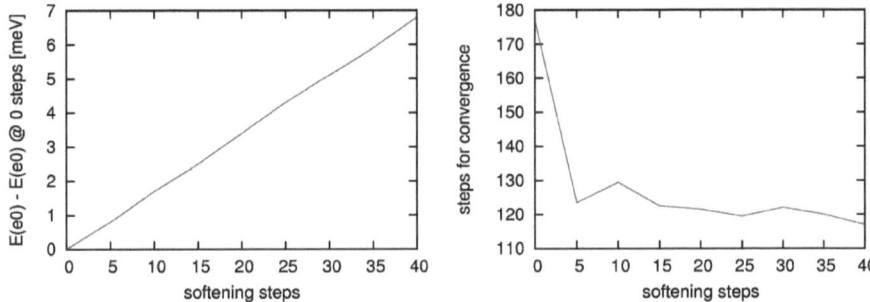

Figure B.2: *First electron state binding energy in a wurtzite GaN quantum dot in an AlN matrix as a function of mesh softening steps (left). Right: Number of minimisation steps per state required to achieve energy convergence below* $5 \cdot 10^{-10}$ *Hartree.*

B.3 Cell size

Within a plane-wave implementation, an investigated isolated nanostructure needs a sufficiently large cell of the matrix material in order to prevent interaction with periodic images. In order to estimate the effect of periodic images, effective mass calculations have been performed for electron states in a spherical GaN quantum dot in AlN for different sphere diameters d. The cell size has been varied from d to $4 \cdot d$ by adding mesh points in all three mesh directions. This means that the resolution of the sphere itself, i.e. the number of mesh point to represent the sphere, does not change for a bigger cell (see Fig. B.3).

The dot size ranges from $d = 0.5$ nm to $d = 2.0$ nm. It is clearly visible that a cell size of four times the dot diameter provides sufficiently converged energy levels even for structures that are significantly smaller than those investigated in this work. For bigger spheres it is visible that already cell sizes of the double sphere diameter provide converged energy levels. Figures B.4 and B.5 show a similar plot for an InN sphere in GaN and an $In_{0.2}Ga_{0.8}N$ sphere in GaN, respectively. The convergence behaviour is basically the same as in Fig. B.3.

B.4 Cutoff energy

The cutoff energy determines the number of plane waves involved in the description of a wave function and, thus, is a measure for the accuracy of a calculation. A large number of plane waves increases the accuracy of the calculation but leads to an increase of computational effort, correspondingly. Therefore, the cutoff energy is a convergence parameter, the influence of which is to be checked. As an example, the influence of the cutoff energy on the binding energies of electron and hole states in a polar grown wurtzite GaN quantum dot (see

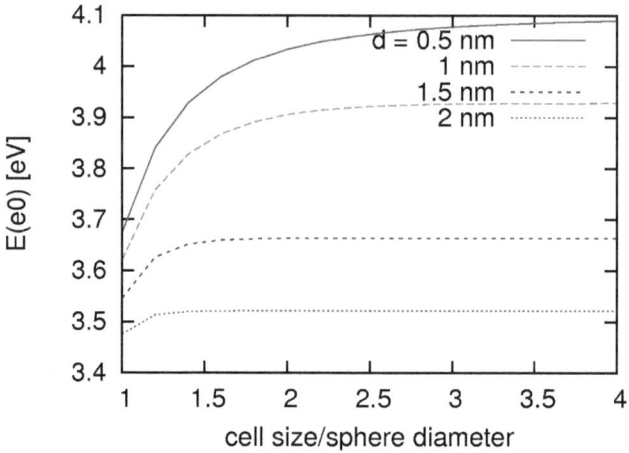

Figure B.3: *Electron ground state binding energy in a spherical zincblende GaN quantum dot in AlN with diameters $d = 0.5$, 1, 1.5 and 2 nm for cell sizes from d to $4 \cdot d$.*

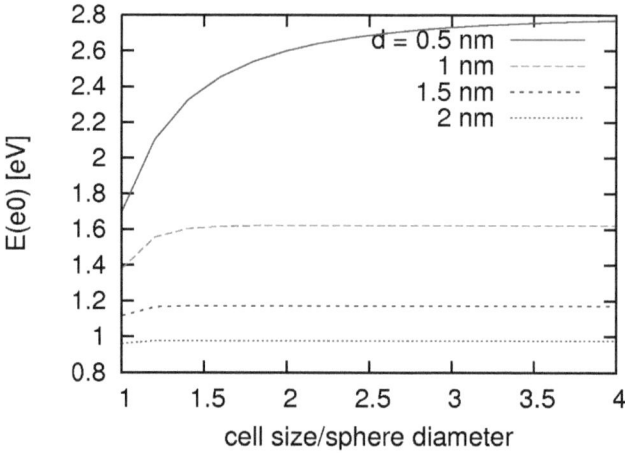

Figure B.4: *Same as Fig. B.3 for an InN sphere in GaN.*

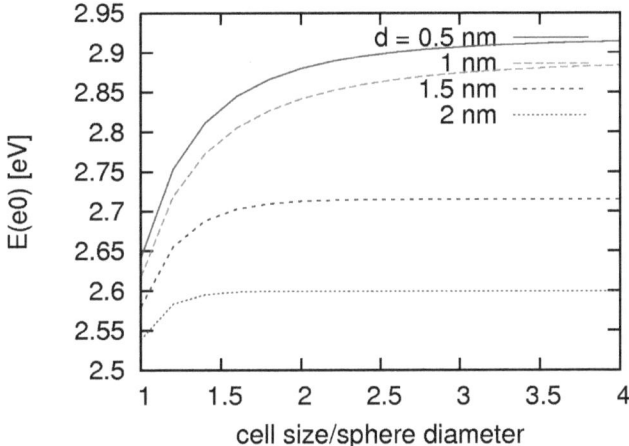

Figure B.5: *Same as Fig.B.3 for an $In_{0.2}Ga_{0.8}N$ sphere in GaN.*

Sec. 4.1.4) is checked. Figure B.6 shows the influence of the cutoff energy on the electron and hole ground state in the studied quantum dot. It can be seen that a cutoff energy above $0.5 \cdot E_{max}^{cut}$ is sufficient to obtain converged eigenvalues. The maximum cutoff energy E_{max}^{cut} [Hartree] is obtained from

$$E_{max}^{cut} = \sqrt{\pi N_i^{mesh}/a_i}, \tag{B.2}$$

where a_i is the cell size along direction i in r_{Bohr} and N_i^{mesh} is the number of mesh points in this direction. This upper limit value is required to prevent wrap-around errors and results from the sampling theorem [32].

B.5 Time and memory consumption

The total time required for the calculation of strain fields, polarisation potentials and each of the eight electron and hole states for a wurtzite InN/GaN quantum dot is given as a function of the Mesh accuracy in Fig. B.7. The electronic states are converged to an accuracy of 10^{-9} H, which corresponds to $2.72 \cdot 10^{-8}$ eV. The cutoff energy is chosen according to Eq. (B.2). The calculation includes the CB-VB coupling, crystal-field and spin-orbit splitting and applies therefore the full eight band model. For a mesh accuracy of $80 \times 80 \times 80$ mesh points, the calculations can be performed in less than 10 h on a standard single processor PC. For this representative example, the calculation of the strain fields and the polarisation

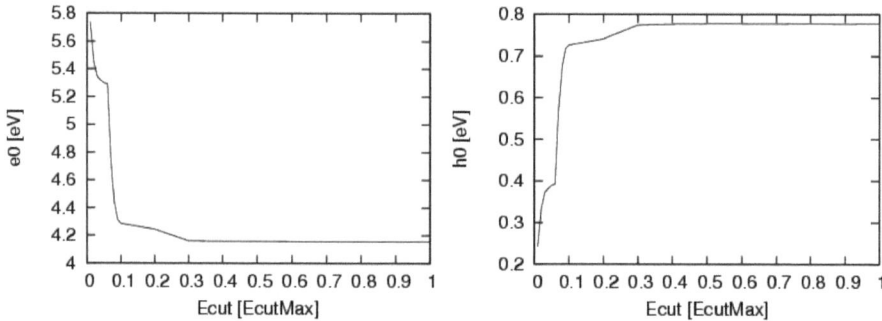

Figure B.6: *First electron (left) and hole (right) state binding energy in a wurtzite GaN quantum dot in an AlN matrix as a function of the cutoff energy E^{cut}.*

potential takes less than 10% of the total time. The calculation of electron and hole states (each eight states) takes approx. 25% (electrons) and 65% (holes) of the total time.

The maximum memory consumption is given for the same calculations in Fig. B.8. This maximum occurs in the calculation of the electronic states. It can be seen, that the memory consumption increases almost linearly with the total number of mesh points and that for a number of $80 \times 80 \times 80$ mesh points less than 2 GB of RAM are required.

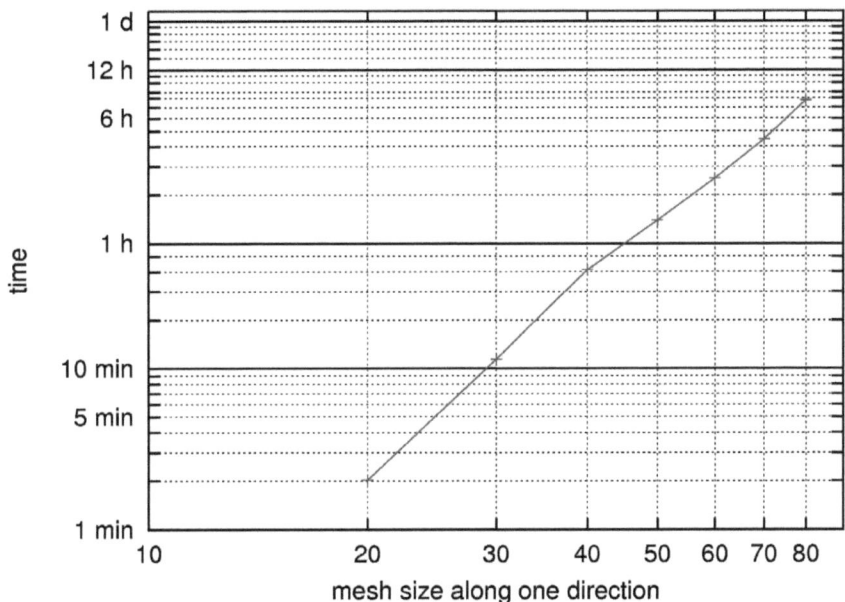

Figure B.7: *Total time for strain field, polarisation potential and electronic structure calculation as a function of mesh points N_r along one direction. The total number of mesh points is $N = N_r^3$.*

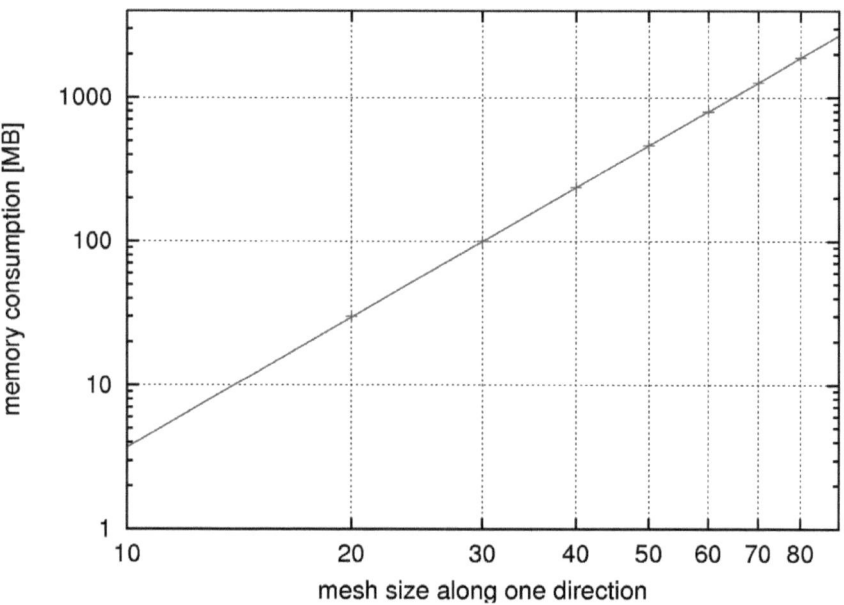

Figure B.8: *Maximum memory consumption as a function of the mesh accuracy.*

Appendix C

Fitting of k · p parameters to a given band structure

The following mathematica script illustrates how **k · p** parameters required for the calculations performed in this work can be derived from a given band structure.
The script fits the parameters A_1 to A_6, $m_e^\perp$ and $m_e^\parallel$ for the wurtzite Hamiltonian in Appendix A to a given band structure (filename *inputBandStructure.dat*). The minimisation is performed for a set of nK **k**-vectors.

$A1 := 0.5/mp$;
$A2 := 0.5/ms$;
$S[\{kx\_, ky\_, kz\_\}] := Ec + A2 * (kx^2 + ky^2) + A1 * kz^2$;
$lambda[\{kx\_, ky\_, kz\_\}] := Ev + 0.5a1 * kz^2 + 0.5a2 * (kx^2 + ky^2)$;
$theta[\{kx\_, ky\_, kz\_\}] := 0.5a3 * kz^2 + 0.5a4 * (kx^2 + ky^2)$;
$F[\{kx\_, ky\_, kz\_\}] := d1 + d2 + lambda[\{kx, ky, kz\}] + theta[\{kx, ky, kz\}]$;
$G[\{kx\_, ky\_, kz\_\}] := d1 - d2 + lambda[\{kx, ky, kz\}] + theta[\{kx, ky, kz\}]$;
$K[\{kx\_, ky\_, kz\_\}] := 0.5a5 * (kx + I * ky)^2$;
$L[\{kx\_, ky\_, kz\_\}] := 0.5a6 * kz * (kx + I * ky)$;

$H[k\_] := \{\{(S[k]), (0), (0), (0), (0), (0), (0), (0)\}$,
$\{(0), (S[k]), (0), (0), (0), (0), (0), (0)\}$,
$\{(0), (0), (F[k]), (-\text{Conjugate}[L[k]]), (-\text{Conjugate}[k[k]]), (0), (0), (0)\}$,
$\{(0), (0), (-L[k]), (lambda[k]), (\text{Conjugate}[L[k]]), (0), (0), (0)\}$,
$\{(0), (0), (-K[k]), (L[k]), (G[k]), (0), (0), (0)\}$,
$\{(0), (0), (0), (0), (0), (G[k]), (-\text{Conjugate}[L[k]]), (-\text{Conjugate}[k[k]])\}$,
$\{(0), (0), (0), (0), (0), (-L[k]), (lambda[k]), (\text{Conjugate}[L[k]])\}$,
$\{(0), (0), (0), (0), (0), (-K[k]), (L[k]), (F[k])\}$
$\}$;

$loadData = \text{Import}["inputBandStructure.dat", "Table"]$;

$eigenvals[k\_] = H[k]//\text{Eigenvalues}$;
$fitFct[kx\_, ky\_, kz\_] := eigenvals[\{kx, ky, kz\}]$;
$SqrDiff = \text{Sum}[\text{Sum}[(loadData[[k, l+3]]$
$- fitFct[loadData[[k, 1]], loadData[[k, 2]], loadData[[k, 3]]][[l]])^2, \{l, 1, 8\}], \{k, 1, nK\}]$;
$\text{Minimize}[\text{Abs}[SqrDiff], \{a1, a2, a3, a4, a5, a6, ms, mp\}]$

Publications list

Parts of this work have been published:

1. O. Marquardt, D. Mourad, S. Schulz, T. Hickel, G. Czycholl and J. Neugebauer: *Comparison of atomistic and continuum theoretical approaches to determine electronic properties of GaN/AlN quantum dots*, Phys. Rev. B **78**, 235302 (2008)

2. T. D. Young and O. Marquardt: *Influence of strain and polarisation on electronic properties of a GaN/AlN quantum dot*, Phys. Stat. Solidi (c) **6**, 557 (2009)

3. O. Marquardt, T. Hickel and J. Neugebauer: *Polarisation-induced charge carrier separation in polar and nonpolar grown GaN quantum dots*, J. Appl. Phys. **106**, 083707 (2009)

4. O. Marquardt, S. Boeck, C. Freysoldt, T. Hickel and J. Neugebauer: *Plane-wave implementation of the real-space $\mathbf{k}\cdot\mathbf{p}$ formalism and continuum elasticity theory*, Computer Phys. Commun. **181**, 765 (2010)

5. H. Abu-Farsakh, O. Marquardt, T. Hickel, L. Lymperakis, and J. Neugebauer: *Theoretical modeling of growth processes, extended defects, and electronic properties of III-nitride semiconductor nanostructures*, submitted to Phys. Stat. Solidi (c)

Acknowledgements

The present investigations were performed within the past four years with the support and help of many different persons. First of all, I wish to thank Prof. Dr. Jörg Neugebauer as well as Dr. Tilmann Hickel for their excellent, inspiring and friendly supervision and for numerous helpful discussions which allowed me to learn and benefit from their impressive knowledge and experience. Their amicable support always created a nice working atmosphere such that I really enjoyed the time of my PhD studies.

For this permanently inspiring and friendly working atmosphere, I furthermore wish to extend my gratitude to my colleagues at the Computational Materials Department and my research partners. In particular, I thank Dr. Sixten Boeck and Dr. Christoph Freysoldt for notable and patient support in the code development within the S/PHI/nX software library and Dr. Liverios Lymperakis for the various nice and instructive discussions about quite a number of different semiconductor-related topics. For the interesting, pleasant and fruitful discussions I furthermore want to thank Dr. Paul Gartner and Dr. Stefan Schulz and our other research partners at Bremen University as well as Prof. Dr. Chris G. Van de Walle (UCSB) and Dr. Toby D. Young (IPPT Warsaw).

Additionally, I would like to express my gratitude to Stefan Kirsch, Andreas Tillack, Dr. Jörg Röseler and Prof. Dr. Burkhard Priemer, who had a crucial influence on my decision for a scientific career in physics. I also like to thank Ugur Aydin for various billard sessions filled with inspiring discussions about physics and non-physics related topics and his help with mathematica.

Last but not least, I would like to thank my parents Marion and Günther Marquardt for constantly encouraging and supporting me on my way.

Bibliography

[1] C. Adelmann, E. Martinez-Guerrero, F. Chabuel, J. Simon, B. Bataillou, G. Mula, L. S. Dang, N. T. Pelekanos, B. Daudin, G. Feuillet and H. Mariette: *Growth and characterisation of self-assembled cubic* GaN *quantum dots*, Materials Science and Engineering B **82**, 212 (2001)

[2] D. Ahn and S. L. Chuang: *Exact calculations of quasibound states of an isolated quantum well with uniform electric field: Quantum-well stark resonance*, Phys. Rev. B **34**, 9034 (1986)

[3] I. Akasaki, S. Sota, H. Sakai, T. Tanaka, M. Koike and H. Amano: *Shortest wavelength semiconductor laser diode*, Electron. Lett. **32**, 1105 (1996)

[4] N. Akopian, G. Bahir, D. Gershoni, M. D. Craven, J. S. Speck and S. P. DenBaars: *Optical evidence for lack of polarisation in ($11\bar{2}0$) oriented* GaN/(AlGa)N *quantum structures*, Appl. Phys. Lett. **86**, 202104 (2005)

[5] M. Albrecht, A. Cremades, J. Krinke, S. Christiansen, O. Ambacher, J. Piqueras, H. P. Strunk and M. Stutzmann: *Carrier Recombination at Screw Dislocations in n-Type* AlGaN *Layers*, phys. stat. sol (b) **216**, 409 (1999)

[6] B. Amstatt, J. Renard, C. Bougerol, E. Bellet-Amalric, B. Gayral and B. Daudin: *Growth of m-plane* GaN *quantum wires and quantum dots on m-plane* 6H-SiC, J. Appl. Phys. **102**, 074913 (2007)

[7] A. D. Andreev and E. P. O'Reilly: *Theory of the electronic structure of* GaN/AlN *hexagonal quantum dots*, Phys. Rev. B **62**, 15851 (2000)

[8] M. Arlery, J. L. Rouvière, F. Widmann, B. Daudin, G. Feuillet and H. Mariette: *Quantitative characterisation of* GaN *quantum-dot structures in* AlN *by high-resolution transmission electron microscopy*, Appl. Phys. Lett. **74**, 3287 (1999)

[9] R. J. Asaro and W. A. Tiller: *Interface morphology development during stress corrosion cracking: Part I. Via surface diffusion*, Metall. Trans. **3**, 1789 (1972)

[10] W. G. Aulbur, L. Jönsson and J. W. Wilkins: *Quasiparticle calculations in solids*, Solid State Phys. **54**, 1 (2000)

[11] N. Baer, P. Gartner and F. Jahnke: *Coulomb effects in semiconductor quantum dots*, Eur. Phys. J. B **42**, 231 (2004)

[12] N. Baer, S. Schulz, P. Gartner, S. Schumacher, G. Czycholl and F. Jahnke: *Influence of symmetry and Coulomb correlation effects on the optical properties of nitride quantum dots*, Phys. Rev. B **76**, 075310 (2007)

[13] T. B. Bahder: *eight band* $\mathbf{k} \cdot \mathbf{p}$ *model of strained zinc-blende crystals*, Phys. Rev. B **41**, 11992 (1990)

[14] T. B. Bahder, R. L. Tober and J. D. Bruno: *Pyroelectric Effect in Semiconductor Heterostructures*, Superlattices Microstruct. **14**, 149 (1993)

[15] T. B. Bahder, R. L. Tober and J. D. Bruno: *Temperature-dependent polarisation in* [111]$In_x Ga_{1-x}As$-$Al_x Ga_{1-x}As$ *quantum wells*, Phys. Rev. B **50**, 2731 (1994)

[16] P. Ballet, P. Disseix, J. Leymarie, A. Vasson, A. M. Vasson and R. Grey: *The determination of* e_{14} *in (111)B-grown* (In,Ga)As/GaAs *strained lasers*, Thin Solid Films **336**, 354 (1998)

[17] J. Bardeen and W. Shockley: *Deformation Potentials and Mobilities in Non-Polar Crystals*, Phys. Rev. **80**, 72 (1950)

[18] A. Barenco, D. Deutsch and A. Ekert: *Conditional Quantum Dynamics and Logic Gates*, Phys. Rev. Lett. **74**, 4083 (1995)

[19] A. Barenco, C. H. Bennett, R. Cleve, D. P. DiVincenzo, N. Margolus, P. Shor, T. Sleator, J. A. Smolin and H. Weinfurter: *Elementary gates for quantum computation*, Phys. Rev. A **25**, 3457 (1995)

[20] G. Bastard: *Superlattice band structure in the envelope-function approximation*, Phys. Rev. B **24**, 5693 (1981)

[21] G. Bastard: *Wave mechanics applied to semiconductor heterostructures* (Les Ulis, 1988)

[22] N. S. Beattie, B. E. Kardynał, A. J. Shields, I. Farrer, D. A. Ritchie and M. Pepper: *Single-photon detection mechanism in a quantum dot transistor*, Physica E **26**, 356 (2004)

[23] I. Belabbas, J. Chen, M. Akli Belkhir, P. Ruterana and G. Nouet: *New core configurations of the c-edge dislocation in wurtzite* GaN, Phys. Stat. Solidi (c) **3**, 1733 (2006)

[24] F. Bernardini, V. Fiorentini and D. Vanderbilt: *Spontaneous polarisation and piezoelectric constants of III-V nitrides*, Phys. Rev. B **56**, R10024 (1997)

[25] F. Bernardini and V. Fiorentini: *Polarisation fields in nitride nanostructures: 10 points to think about*, Appl. Surf. Sci. **166**, 23 (2000)

[26] F. Bernardini, V. Fiorentini and D. Vanderbilt: *Accurate calculation of polarisation-related quantities in semiconductors*, Phys. Rev. B **63**, 193201 (2001)

[27] F. Bernardini and V. Fiorentini: *Nonlinear macroscopic polarisation in III-V nitride alloys*, Phys. Rev. B **64**, 085207 (2001)

[28] F. Bernardini and V. Fiorentini: *First-principles calculation of the piezoelectric tensor d of III-V nitrides*, Appl. Phys. Lett. **80**, 4145 (2002)

[29] G. Bester and A. Zunger: *Cylindrically shaped zinc-blende semiconductor quantum dots do not have cylindrical symmetry:Atomistic symmetry, atomic relaxation, and piezoelectric effects*, Phys. Rev. B **71**, 045318 (2005)

[30] E. Biolatti, R. C. Iotti, P. Zanardi and F. Rossi: *Quantum Information Processing with Semiconductor Macroatoms*, Phys. Rev. Lett. **85**, 5647 (2000)

[31] G. L. Bir and G. E. Pikus: *Symmetry and Strain-Induced Effects in Semiconductors* (Wiley, New York, 1975)

[32] S. Boeck: *Development and Application of the S/PHI/nX Library: First-principles Calculations of Thermodynamic Properties of III-V Semiconductors*, PhD thesis, Paderborn (2009)

[33] D. Bouwmeester, A. Ekert and A. Zeilinger, eds.: *The Physics of Quantum Information* (Springer, Berlin, 2000)

[34] T. Bretagnon, P. Lefebvre, P. Valvin, R. Bardoux, T. Guillet, T. Taliercio, B. Gil, N. Grandjean, F. Semond, B. Damilano, A. Dussaigne and J. Massies: *Radiative lifetime of a single electron-hole pair in GaN/AlN quantum dots*, Phys. Rev. B **73**, 113304 (2006)

[35] K. Brugger: *Thermodynamic Definition of Higher Order Elastic Coefficients*, Phys. Rev. **133**, A1611 (1964)

[36] G. Bu, D. Ciplys, M. Shur, L. J. Schowalter, S. Schujman and R. Gaska: *Electromechanical coupling coefficient for surface acoustic waves in single-crystal bulk aluminum nitride*, Appl. Phys. Lett. **84**, 4611 (2004)

[37] M. G. Burt: *The justification for applying the effective-mass approximation to microstructures*, J. Phys. Condens. Matter. **4**, 6651 (1992)

[38] M. G. Burt: *Fundamentals of envelope function theory for electronic states and photonic modes in nanostructures*, J. Phys.: Condens. Matter **11** R53 (1999)

[39] A. D. Bykhovski, V. V. Kaminski, M. S. Shur, Q. C. Chen and M. A. Khan: *Piezoresistive effect in wurtzite n-type GaN*, Appl. Phys. Lett. **68**, 818 (1996)

[40] W. D. Callister, Jr.: *Fundamentals of Materials Science and Engineering* (John Wiley and Sons, Danvers, MA. 2005)

[41] M. Cardona and F. H. Pollack: *Energy-Band Structure of Germanium and Silicon: The* $\mathbf{k} \cdot \mathbf{p}$ *Method*, Phys. Rev. **142**, 530 (1966)

[42] D. M. Ceperley and B. J. Alder: *Ground State of the Electron Gas by a Stochastic Method*, Phys. Rev. Lett. **45**, 566 (1980)

[43] Y. C. Chang: *Bond-orbital models for superlattices*, Phys. Rev. B **37**, 8215 (1988)

[44] J. R. Chelikowsky and M. L. Cohen: *Electronic structure of silicon*, Phys. Rev. **10**, 5095 (1974)

[45] J. R. Chelikowsky and S. G. Louie: *Quantum Theory of Real Materials* (Kluwer Academic, Dordrecht, The Netherlands, 1996)

[46] S. F. Chichibu, T. Azuhata, T. Sota and S. Nakamura: *Spontaneous emission of localized excitons in* InGaN *single multiquantum well structures*, Appl. Phys. Lett. **69**, 30 (1996)

[47] S. F. Chichibu, T. Azuhata, H. Okumura, A. Tackeuchi, T. Sota and T. Mukai: *Localized exciton dynamics in* InGaN *quantum well structures*, Appl. Surf. Sci. **190**, 330 (2002)

[48] A. Chitnis, C. Chen, V. Adivarahan, M. Shatalov, E. Kuokstis, V. Mandavilli, J. Yang and M. A. Khan: *Visible light-emitting diodes using a-plane* GaN/InGaN *multiple quantum wells over r-plane sapphire*, Appl. Phys. Lett. **84**, 3663 (2004)

[49] C. H. Chiu, S. Y. Kuo, M. H. Lo, C. C. Ke, T. C. Wang, Y. T. Lee, H. C. Kuo, T. C. Lu and S. C. Wang: *Optical properties of a-plane* InGaN/GaN *multiple quantum wells on r-plane sapphire substrates with different indium compositions*, J. Appl. Phys. **105**, 063105 (2009)

[50] A. Y. Cho and J. R. Arthur: *Molecular beam epitaxy*, Prog. Solid State Chem. **10**, 157 (1975)

[51] S. Cho, J. Kim, A. Sanz-Hervás, A. Majerfeld, G. Patriarche and B. W. Kim: *Characterization of piezoelectric and pyroelectric properties of MOVPE-grown strained (111)A* InGaAs/GaAs *QW structures by modulation spectroscopy*, Phys. Stat. Solidi (a) **195**, 260 (2003)

[52] H. J. Choi, J. C. Johnson, R. He, S. K. Lee, F. Kim, P. Pauzauskie, J. Goldberger, R. J. Saykally and P. Yang: *Self-Organized* GaN *Quantum Wire UV Lasers*, J. Phys. Chem. B, **107**, 8721 (2003)

[53] S. L. Chuang and C. S. Chang: $\mathbf{k} \cdot \mathbf{p}$ *method for strained wurtzite semiconductors*, Phys. Rev. B **54**, 2491 (1996)

[54] M. L. Cohen and J. R. Chelikowsky: *Electronic Structure and Optical Properties of Semiconductors*, Solid State Sciences **75** (Springer, Berlin, 1988)

[55] T. E. Cook, Jr., C. C. Fulton, W. J. Mecouch, R. F. Davis, G. Lucovsky and R. J. Nemanich: *Band offset measurements of the Si_3N_4/GaN (0001) interface*, J. Appl. Phys. **94**, 3949 (2003)

[56] A. Dal Corso, M. Posternak, R. Resta and A. Baldereschi: *Ab initio study of piezoelectricity and spontaneous polarisation in ZnO*, Phys. Rev. **B** 50, 10715 (1994)

[57] S. Cortez, O. Krebs and P. Voisin: *In-plane optical anisotropy of quantum well structures: From fundamental considerations to interface characterization and optoelectronic engineering*, J. Vac. Sci. Technol. B **18**, 2232 (2000)

[58] M. D. Craven, P. Waltereit, J. S. Speck and S. P. DenBaars: *Well-width dependence of photoluminescence emission from a-plane GaN/AlGaN multiple quantum wells*, Appl. Phys. Lett. **84**, 496 (2004)

[59] A. Cremades and J. Piqueras: *Study of carrier recombination at structural defects in InGaN films*, Mater. Sci. Eng. **B91-92**, 341 (2002)

[60] A. Cros, J. A. Budagosky, A. García-Cristóbal, N. Garro, A. Cantarero, S. Founta, H. Mariette and B. Daudin: *Influence of strain in the reduction of the internal electric field in GaN/AlN quantum dots grown on a-plane 6H-SiC*, Phys. Stat. Solidi (b) **243**, 1499 (2006)

[61] Y. Cui, Z. Zhong, D. Wang, W. U. Wang and C. M. Lieber: *High Performance Silicon Nanowire Field Effect Transistors*, Nano Lett. **3**, 149 (2003)

[62] B. Daudin, G. Feuillet, H. Mariette, G. Mula, N. T. Pelekanos, E. Molva, J. L. Rouvière, C. Adelmann, E. Martinez-Guerrero, J. Barjon, F. Chabuel, B. Bataillou and J. Simon: *Self-Assembled GaN Quantum Dots Grown by Plasma-Assisted Molecular Beam Epitaxy*, Jpn. J. Appl. Phys. **40**, 1892 (2001)

[63] B. Daudin: *Polar and nonpolar GaN quantum dots*, J. Phys.: Condens. Matter **20**, 473201 (2008)

[64] E. Dekel, D. Gershoni, E. Ehrenfreund, D. Spector, J. M. Garcia and P. M. Petroff: *Multiexciton Spectroscopy of a Single Self-Assembled Quantum Dot*, Phys. Rev. Lett. **80**, 4991 (1998)

[65] E. Dekel, D. Gershoni, E. Ehrenfreund, J. M. Garcia and P. M. Petroff: *Carrier-carrier correlations in an optically excited single semiconductor quantum dot*, Phys. Rev. B **61**, 11009 (2000)

[66] D. Deutsch: *Quantum mechanics near closed timelike lines*, Phys. Rev. D **44**, 3197 (1991)

[67] P. Dłużewski, G. Maciejewski, G. Jurczak, S. Kret and J. L. Laval: *Nonlinear FE analysis of residual stresses induced by dislocations in heterostructures*, Comput. Mater. Sci. **29**, 379 (2004)

[68] X. Duan, Y. Huang, R. Agarwal and C. M. Lieber: *Single-nanowire electrically driven lasers*, Nature **421**, 241 (2002)

[69] H. Ehrenreich, F. Seitz and D. Turnbull: *Solid State Physics, Advances and Applications* (Academic Press, New York, 1954)

[70] C. Fiolhais, F. Nogueira and M. Marques: *A Primer in Density Functional Theory* (Springer, Berlin, 2003)

[71] M. E. Flatté, P. M. Young, L.-H. Peng and H. Ehrenreich: *Generalized superlattice* $\mathbf{k} \cdot \mathbf{p}$ *theory and intersubband optical transitions*, Phys. Rev. B **53**, 1963 (1996)

[72] V. A. Fonoberov and A. A. Balandin: *Excitonic properties of strained wurtzite and zincblende* $GaN/Al_xGa_{1-x}N$ *quantum dots*, J. Appl. Phys. **94**, 7178 (2003)

[73] V. A. Fonoberov and A. A. Balandin: *Optical properties of wurtzite and zinc-blende GaN/AlN quantum dots*, J. Vac. Sci. Technol. B **22**, 2190 (2004)

[74] B. A. Foreman: *Effective-mass Hamiltonian and boundary conditions for the valence bands of semiconductor microstructures*, Phys. Rev. B **48**, 4964 (1993)

[75] S. Founta, F. Rol, E. Bellet-Amalric, J. Bleuse, B. Daudin, B. Gayral and H. Mariette: *Optical properties of GaN grown on nonpolar* $(11-20)$ *SiC by molecular-beam epitaxy*, Appl. Phys. Lett. **86**, 171901 (2005)

[76] S. Founta, J. Coraux, D. Jalabert, C. Bougerol, F. Rol, H. Mariette, H. Renevier and B. Daudin: *Anisotropic strain relaxation in a-plane GaN quantum dots*, J. Appl. Phys. **101**, 063541 (2007)

[77] S. Founta, C. Bougerol, H. Mariette, B. Daudin and P. Vennéguès: *Anisotropic morphology of nonpolar a-plane GaN quantum dots and quantum wells*, J. Appl. Phys. **102**, 074304 (2007)

[78] N. Fraj, I. Saïdi, S. Ben Radhia and K. Boujdaria: *Band structures of AlAs, GaP, and SiGe alloys: A 30* $\mathbf{k} \cdot \mathbf{p}$ *model*, J. Appl. Phys. **102**, 053703 (2007)

[79] F. C. Frank and J. H. van der Merwe: *One-Dimensional Dislocations. I. Static Theory*, Proc. Roy. Soc. London A **198**, 205 (1949)

[80] D. Fritsch, H. Schmidt and M. Grundmann:*Band-structure pseudopotential calculation of zincblende and wurtzite AlN, GaN, and InN*, Phys. Rev. B **67**, 235205 (2003)

[81] T. Fukui, S. Ando, T. Honda and T. Toriyama: GaAs *tetrahedral quantum dot structures fabricated using area MOCVD*, Surf. Sci. **267**, 236 (1992)

[82] M. Funato and Y. Kawakami: *Excitonic properties of polar, semipolar and nonpolar InGaN/GaN strained quantum wells with potential fluctuations*, J. Appl. Phys. **103**, 093501 (2008)

[83] J. P. Garayt, J. M. Gerard, F. Enjalbert, L. Ferlazzo, S. Founta, E. Martinez-Guerrero, F. Rol, D. Araujo, R. Cox, B. Daudin, B. Gayral, L. S. Dang and H. Mariette:*Study of isolated cubic GaN quantum dots by low-temperature cathodoluminescence*, Physica E (Amsterdam) **26**, 203 (2005)

[84] N. Garro, A. Cros, J. A. Budagosky, A. Cantarero, A. Vinattieri, M. Gurioli, S. Founta, H. Mariette and B. Daudin: *Reduction of the internal electric field in wurtzite a-plane GaN self-assembled quantum dots*, Appl. Phys. Lett. **87**, 011101 (2005)

[85] L. Geelhaar, C. Chèze, W. M. Weber, R. Averbeck, H. Riechert, T. Kehagias, P. Komninou, G. P. Dimitrakopulos and T. Karakostas: *Axial and radial growth of* Ni-*induced* GaN *nanowires*, Appl. Phys. Lett. **91**, 093113 (2007)

[86] J. M. Gerard and B. Gayral:*Strong Purcell effect for* InAs *quantum boxes in three-dimensional solid-state microcavities*, J. Light Wave Tech. **17**, 2089 (1999)

[87] D. Gershoni, C. Henry and G. Baraff: *Calculating the optical properties of multidimensional heterostructures: Application to the modeling of quaternary quantum well lasers*, IEEE J. Quant. Elec. **29**, 2433 (1993)

[88] N. Gogneau, F. Enjalbert, F. Fossard, Y. Hori, C. Adelmann, J. Brault, E. Martinez-Guerrero, J. Simon, E. Bellet-Amalric, D. Jalabert, N. Pelekanos, J. L. Rouvière, B. Daudin, L. S. Dang, H. Mariette and E. Monroy: *Recent progress in growth and physics of* GaN/AlN *quantum dots*, Phys. Stat. Solidi (c) **1**, 1445 (2004)

[89] I. Goroff and L. Kleinman: *Deformation Potentials in Silicon. III. Effects of a General Strain on Conduction and Valence Levels*, Phys. Rev. **132**, 1080 (1963)

[90] K. Goshima, K. Komori, T. Sugaya and T. Takagahara: *Formation and control of a correlated exciton two-qubit system in a coupled quantum dot*, Phys. Rev. B **79**, 205313 (2009)

[91] A. O. Govorov: *Spin-Förster transfer in optically excited quantum dots*, Phys. Rev. B **71**, 155323 (2005)

[92] N. Grandjean, B. Damilano, S. Dalmasso, M. Leroux, M. Laügt and J. Massies: *Built-in electric-field effects in wurtzite* AlGaN/GaN *quantum wells*, J. Appl. Phys. **86**, 3714 (1999)

[93] M. A. Grinfeld: *Instability of an interface between a nonhydrostatically stressed crystal and a melt*, Dokl. Akad. Nauk. SSSR **290**, 1358 (1986)

[94] G. Grosso, S. Moroni and G. Pastori Parravicini: *Electronic structure of the InAs-GaSb superlattice studied by the renormalization method*, Phys. Rev. B **40**, 12328 (1989)

[95] M. Grundmann, O. Stier and D. Bimberg: *InAs/GaAs pyramidal quantum dots: Strain distribution, optical phonons, and electronic structure*, Phys. Rev. B **52**, 11969 (1995)

[96] P. J. Hansen, Y. E. Strausser, A. N. Erickson, E. J. Tarsa, P. Kozodoy, E. G. Brazel, J. P. Ibbetson, U. Mishra, V. Narayanamurti, S. P. DenBaars and J. S. Speck: *Scanning capacitance microscopy imaging of threading dislocations in GaN films grown on (0001) sapphire by metalorganic chemical vapor deposition*, Appl. Phys. Lett. **72**, 2247 (1998)

[97] H. Hasegawa: *Theory of Cyclotron Resonance in Strained Silicon Crystals*, Phys. Rev. **129**, 1029 (1963)

[98] B. A. Haskell, F. Wu, S. Matsuda, M. D. Craven, P. T. Fini, S. P. Den-Baars, J. S. Speck and S. Nakamura: *Structural and morphological characteristics of planar ($11\bar{2}0$) a-plane gallium nitride grown by hydride vapor phase epitaxy*, Appl. Phys. Lett. **83**, 1554 (2003)

[99] P. Hawrylak: *Excitonic artificial atoms: Engineering optical properties of quantum dots*, Phys. Rev. B **60**, 5597 (1999)

[100] O. Hayden, R. Agarwal and C. M. Lieber: *Nanoscale avalanche photodiodes for highly sensitive and spatially resolved photon detection*, Nat. Mater. **5**, 352 (2006)

[101] L. Hedin: *New Method for Calculating the One-Particle Green's Function with Application to the Electron-Gas Problem*, Phys. Rev. **139**, A796 (1965)

[102] R. Heitz, F. Guffarth, K. Pötschke, A. Schliwa and D. Bimberg: *Shell-like formation of self-organized InAs/GaAs quantum dots*, Phys. Rev. B **71**, 045325 (2005)

[103] S. D. Hersee, X. Sun and X. Wang: *The Controlled Growth of GaN nanowires*, Nano Lett. **6**, 1808 (2006)

[104] T. Hino, S. Tomiya, T. Miyajima, K. Yanashima, S. Hashimoto and M. Ikeda: *Characterization of threading dislocations in GaN epitaxial layers*, Appl. Phys. Lett. **76**, 3421 (2000)

[105] I.h. Ho and G. B. Stringfellow: *Solid phase immiscibility in GaInN*, Appl. Phys. Lett. **64**, 1687 (1996)

[106] P. Hohenberg and W. Kohn: *Inhomogeneous Electron Gas*, Phys. Rev. **136**, B864 (1964)

[107] K. Hoshino and Y. Arakawa: *UV photoluminescence from GaN self-assembled quantum dots on $Al_xGa_{1-x}N$ surfaces grown by metalorganic chemical vapor deposition*, Phys. Stat. Solidi (c) **1**, 2516 (2004)

[108] Y. Huang, X. Duan, Y. Cui, L. J. Lauhon, K. H. Kim and C. M. Lieber: *Logic Gates and Computation from Assembled Nanowire Building Blocks*, Science **294**, 1313 (2001)

[109] L. Jacak, P. Hawrylak and A. Wójs: *Quantum Dots* (Springer, Berlin, 1998)

[110] K. Jacobi: *Atomic structure of* InAs *quantum dots on* GaAs, Prog. Surf. Sci. **71**, 185 (2003)

[111] H. Jiang and J. Singh: *Strain distribution and electronic spectra of* InAs/GaAs *self-assembled dots: an eight band study*, Phys. Rev. B **56**, 4696 (1997)

[112] B. Jogai: *Three-dimensional strain field calculations in coupled* InAs/GaAs *quantum dots*, J. Appl. Phys. **88**, 5050 (2000)

[113] B. Jogai: *Three-dimensional strain field calculations in multiple* InN/AlN *wurtzite quantum dots*, J. Appl. Phys. **90**, 699 (2001)

[114] R. Jones: *Do we really understand dislocations in semiconductors?*, Mater. Sci. Eng. **B71**, 24 (2000)

[115] E. O. Kane: *Energy Band Structure in p-type Germanium and Silicon*, J. Phys. Chem. Solids **1**, 82 (1956)

[116] E. O. Kane, in: *Handbook on semiconductors vol. 1* (Elsevier, Amsterdam, 1982)

[117] E. Kapon: *Semiconductor Lasers I: Fundamentals* (Academic Press, San Diego, 1999)

[118] K. Kawamoto, T. Suda, T. Akiyama, K. Nakamura and T. Ito: *An empirical potential approach to dislocation formation and structural stability in* GaN_xAs_{1-x}, Appl. Surf. Sci. **244**, 182 (2005)

[119] K. Kawasaki, D. Yamazaki, A. Kinoshita, H. Hirayama, K. Tsutsui and Y. Aoyagi: GaN *quantum-dot formation by self-assembling droplet epitaxy and application to single-electron transistors*, Appl. Phys. Lett. **79**, 2243 (2001)

[120] P. N. Keating: *Effect of Invariance Requirements on the Elastic Strain Energy of Crystals with Application to the Diamond Structure*, Phys. Rev. **145**, 637 (1996)

[121] K. C. Kim, M. C. Smith, H. Sato, F. Wu, N. Fellows, M. Saito, K. Fujito, J. S. Speck, S. Nakamura and S. P. DenBaars: *Improved electroluminescence on nonpolar m-plane* InGaN/GaN *quantum wells LEDs*, Phys. Stat. Solidi (RPL) **1**, 125 (2007)

[122] R. D. King-Smith and D. Vanderbilt: *Theory of polarisation of crystalline solids*, Phys. Rev. B **47**, 1651 (1993)

[123] W. H. Kleiner and M. Roth: *Deformation Potential in Germanium from Optical Absorption Lines for Exciton Formation*, Phys. Rev. Lett. **2**, 334 (1959)

[124] V. I. Klimov, A. A. Mikhailovsky, S. Xu, A. Malko, J. A. Hollingsworth, C. A. Leatherdale, H. J. Eisler and M. G. Bawendi: *Optical Gain and Stimulated Emission in Nanocrystal Quantum Dots*, Science **290**, 314 (2000)

[125] W. Kohn and L. J. Sham: *Self-Consistent Equation including Exchange and Correlation Effects*, Phys. Rev. **140**, A1133 (1965)

[126] K. Kojima, H. Kamon, M. Funato and Y. Kawakami: *Theoretical investigations on anisotropic optical properties in semipolar and nonpolar InGaN quantum wells*, Phys. Stat. Solidi (c) **5**, 3038 (2008)

[127] Y. Kokubun, J. Nishio, M. Abe, T. Ehara and S. Nakagomi: *Properties of GaN Epitaxial Layers grown ad high Growth Rates by Metalorganic Chemical Vapor Deposition*, J. Electron. Mat. **30**, 23 (2001)

[128] S. Komiyama, O. Astafiev, V. Antonov, H. Hirai and T. Kutsuwa: *A single-photon detector in the far-infrared range*, Nature **403**, 405 (2000)

[129] M. Kneissl, D. Treat, M. Teepe, N. Miyashita and N. M. Johnson: *Continuous-wave operation of ultraviolet InGaN/InAlGaN multiple-quantum-well laser diodes*, Appl. Phys. Lett. **82**, 2386 (2003)

[130] M. El Kurdi, G. Fishman, S. Sauvage, P. Boucaud: *Comparison between 6-band and 14-band* **k** · **p** *formalisms in SiGe/Si heterostructures*, Phys. Rev. B **68**, 165333 (2003)

[131] Y. H. Kwon, G. H. Gainer, S. Bidnyk, Y. H. Cho, J. J. Song, M. Hansen and S. P. DenBaars: *Structural and optical characteristics of* $In_xGa_{1-x}N$/GaN *multiple quantum wells with different In compositions*, Appl. Phys. Lett. **75**, 2545 (1999)

[132] C. Lang, D. Nguyen-Manh and D. J. H. Cockayne: *Nonuniform alloying in Ge(Si)/Si(001) quantum dots*, J. Appl. Phys. **94**, 7067 (2003)

[133] L. Lari, R. T. Murray, M. H. Gass, T. J. Bullough, P. R. Chalker, J. Kioseoglou, G. P. Dimitrakopulos, T. Kehagias, P. Komninou, T. Karakostas, C. Chèze, L. Geelhaar and H. Riechert: *Defect characterization and analysis of III-V nanowires grown by Ni-promoted MBE*, Phys. Stat. Solidi (a) **205**, 2589 (2008)

[134] L. Lari, R. T. Murray, T. J. Bullough, P. R. Chalker, M. H. Gass, C. Chèze, L. Geelhaar and H. Riechert: *Nanoscale compositional analysis of Ni-based seed crystallites associated with GaN nanowire growth*, Physica E **40**, 2457 (2008)

[135] L. Lari, R. T. Murray, M. H. Gass, T. J. Bullough, P. R. Chalker, C. Chèze, L. Geelhaar and H. Riechert: *Nanoscale EELS and EDX Analyses of GaN Nanowires and GaN/AlGaN Radial Heterostructure Nanowires*, Microsc. Microanal. **14**, 1394 (2008)

[136] S. Lazar, C. Hébert and H. Zandbergen: *Investigation of hexagonal and cubic GaN by high-resolution electron energy-loss spectroscopy and density functional theory*, Ultramicroscopy **98**, 249 (2004)

[137] N. N. Ledentsov, V. M. Ustinov, V. A. Shchukin, P. S. Kopev, Zh. I. Alferov and D. Bimberg: *Quantum dot heterostructures: Fabrication, properties, lasers*, Semiconductors **32**, 343 (1998)

[138] Z. Y. Li, M. H. Lo, C. H. Chiu, P. C. Lin, T. C. Lu, H. C. Kuo and S. C. Wang: *Carrier localisation degree of $In_{0.2}Ga_{0.8}N/GaN$ multiple quantum wells grown on vicinal sapphire substrates*, J. Appl. Phys. **105**, 013103 (2009)

[139] C. M. Lieber and Z. L. Wang: *Functional Nanowires*, MRS Bull. **32**, 99 (2007)

[140] Z. Liliental-Weber, Y. Chen, S. Ruvimov and J. Washburn: *Formation Mechanism of Nanotubes in GaN*, Phys. Rev. Lett. **79**, 2835 (1997)

[141] Z. Liliental-Weber, J. Jasinski, J. Washburn and M. A. O'Keefe: *Screw dislocations in GaN*, 60th Ann. Proc. MSA, Quebec, Canada (2002)

[142] B. Liu, R. Zhang, Z. L. Xie, C. X. Liu, J. Y. Kong, J. Yao, Q. J. Liu, Z. Zhang, D. Y. Fu, X. Q. Xiu, H. Lu, P. Chen, P. Han, S. L. Gu, Y. Shi and Y. D. Zheng: *Nonpolar m-plane thin film GaN and InGaN/GaN light emitting diodes on $LiAlO_2(100)$ substrates*, Appl. Phys. Lett. **91**, 253506 (2007)

[143] J. P. Loehr: *Improved effective-bond-orbital model for superlattices*, Phys. Rev. B **50**, 5429 (1994)

[144] J. P. Loehr: *Physics of strained quantum well lasers* (Kluwer, Boston, 1998)

[145] H. Lu, W. J. Schaff, L. F. Eastman, J. Wu, W. Walukiewicz, V. Cimalla and O. Ambacher: *Growth of a-plane InN on r-plane sapphire with a GaN buffer by molecular-beam epitaxy*, Appl. Phys. Lett. **83**, 1136 (2003)

[146] J. M. Luttinger and W. Kohn: *Motion of Electrons and Holes in Perturbed Periodic Fields*, Phys. Rev. **97**, 869 (1955)

[147] L. Lymperakis, J. Neugebauer, M. Albrecht, T. Remmele and H. P. Strunk: *Strain Induced Deep Electronic States around Threading Dislocations in GaN*, Phys. Rev. Lett. **93**, 196401 (2004)

[148] O. Madelung: *Introduction to Solid-State Theory* (Springer, Berlin, Heidelberg, 1978)

[149] H. M. Manasevit: *Single-Crystal Gallium Arsenide on Insulating Substrates*, Appl. Phys. Lett. **12**, 156 (1968)

[150] E. Martinez-Guerrero, C. Adelmann, F. Chabuel, J. Simon, N. T. Pelekanos, G. Mula, B. Daudin, G. Feuillet and H. Mariette: *Self-assembled zinc blende GaN quantum dots grown by molecular-beam epitaxy*, Appl. Phys. Lett. **77**, 809 (2000)

[151] O. Marquardt, D. Mourad, S. Schulz, T. Hickel, G. Czycholl and J. Neugebauer: *Comparison of atomistic and continuum theoretical approaches to determine electronic properties of GaN/AlN quantum dots*, Phys. Rev. B **78**, 235302 (2008)

[152] O. Marquardt, T. Hickel and J. Neugebauer: *Polarisation-induced charge carrier separation in polar and nonpolar grown GaN quantum dots*, J. Appl. Phys. **106**, 083707 (2009)

[153] O. Marquardt, S. Boeck, C. Freysoldt, T. Hickel and J. Neugebauer: *Plane-wave implementation of the real-space $\mathbf{k} \cdot \mathbf{p}$ formalism and continuum elasticity theory*, Computer. Phys. Commun. **181**, 765 (2010)

[154] N. Marzari and D. Vanderbilt: *Maximally localized generalized Wannier functions for composite energy bands*, Phys. Rev. B **56**, 12847 (1997)

[155] J. Marzin and G. Bastard: *Calculation of the energy levels in InAs/GaAs quantum dots*, Solid State Commun. **92**, 437 (1994)

[156] M. D. McCluskey, C. G. Van de Walle, C. P. Master, L. T. Romano and N. M. Johnson: *Large band gap bowing of $In_xGa_{1-x}N$ alloys*, Appl. Phys. Lett. **72**, 2725 (1998)

[157] Y. Mera and K. Maeda: *Optoelectronic Activities of Dislocations in Gallium Nitride Crystals*, IEICE Trans. Electron. **E83-C**, 612 (2000)

[158] S. Miasojedovas, S. Juršėnas, G. Kurilčik, A. Žukauskas, M. Springis, I. Tale and C. C. Yang: *Stimulated Emission in InGaN/GaN Multiple Quantum Wells with Different Indium Content*, Acta Phys. Pol. A **107**, 256 (2005)

[159] P. Michler, A. Kiraz, C. Becher, W. V. Schoenfeld, P. M. Petroff, L. Zhang, E. Hu and A. Imamoğlu: *A quantum dot single-photon turnstyle device*, Science **290**, 2282 (2000)

[160] D. A. B. Miller, D. S. Chemla, T. C. Damen, A. C. Gossard, W. Wiegmann, T. H. Wood and C. A. Burrus: *Band-Edge Electroabsorption in Quantum Well Structures: The Quantum-Confined Stark Effect*, Phys. Rev. Lett. **53**, 2173 (1984)

[161] S. Muensit, E. M. Goldys and I. L. Guy: *Shear piezoelectric coefficients of gallium nitride and aluminum nitride*, Appl. Phys. Lett. **75**, 3965 (1999)

[162] S. Nagahama, T. Yanamoto, M. Sano and T. Mukai: *Characteristics of Ultraviolet Laser Diodes Composed of Quaternary $Al_xIn_yGa_{(1-x-y)}N$*, Jpn. J. Appl. Phys. **40**, L788 (2001)

[163] S. Nagahama, T. Yanamoto, M. Sano and T. Mukai: *Study of GaN-based Laser Diodes in Near Ultraviolet Region*, Jpn. J. Appl. Phys., Part 1 **41**, 5 (2002)

[164] S. Nakamura, M. Senoh, N. Iwasa, S. Nagahama, T. Yamada and T. Mukai: *Superbright Green InGaN Single-Quantum-Well-Structure Light-Emitting Diodes*, Jpn. J. Appl. Phys. **34**, L1332 (1995)

[165] S. Nakamura, M. Senoh, S. Nagahama, N. Iwasa, T. Yamada, T. Matsushita, H. Kiyoku and Y. Sugimoto: *InGaN-Based Multi-Quantum-Well-Structure Laser Diodes*, Jpn. J. Appl. Phys. **35**, L74 (1996)

[166] S. Nakamura, M. Senoh, S. Nagahama, N. Iwasa, T. Yamada, T. Matsushita, H. Kiyoku and Y. Sugimoto: *InGaN multi-quantum-well structure laser diodes grown on* $MgAl_2O_4$ *substrates*, Appl. Phys. Lett. **68**, 2105 (1996)

[167] S. Nakamura and G. Fasol: *The Blue Laser Diode* (Springer, Berlin, 1997)

[168] M. B. Nardelli, K. Rapcewicz and J. Bernholc: *Polarisation field effects on the electron-hole recombination dynamics in* $In_{0.2}Ga_{0.8}N/In_{1-x}Ga_xN$ *multiple quantum wells*, Appl. Phys. Lett. **71**, 3135 (1997)

[169] Y. Narukawa, Y. Kawakami, M. Funato, Shizuo Fujita, Shigeo Fujita and S. Nakamura: *Role of self-formed InGaN quantum dots for exciton localisation in the purple laser diode emitting at 420 nm*, Appl. Phys. Lett. **70**, 981 (1997)

[170] M. A. Nelsen and I. L. Chuang: *Quantum Computation and Quantum Information* (Cambridge University Press, Cambridge, England, 2000)

[171] H. M. Ng, A. Bell, F. A. Ponce and S. N. G. Chu: *Structural and optical characterization of nonpolar GaN/AlN quantum wells*, Appl. Phys. Lett. **83**, 653 (2003)

[172] R. Nötzel: *Self-organized growth of quantum-dot structures*, Semicond. Sci. Technol. **11**, 1365 (1996)

[173] J. H. Noh, H. Asahi, S. J. Kim and S. Gonda: *Scanning Tunneling Microscopy/Spectroscopy Study of Self-Organized Quantum Dot Structures Formed in GaP/InP Short-Period Superlattices*, Jpn. J. Appl. Phys. **36**, 3818 (1997)

[174] P. Nozieres: *Theory of Interacting Fermi Systems* (Benjamin, New York, 1964)

[175] K. P. O'Donnell, R. W. Martin and P. G. Middleton: *Origin of Luminescence from InGaN Diodes*, Phys. Rev. Lett. **82**, 237 (1999)

[176] G. Onida, L. Reining and A. Rubio: *Electronic excitations: density-functional versus many-body Greens-function approaches*, Rev. Mod. Phys. **74**, 601 (2002)

[177] N. Onojima, J. Suda and H. Matsunami: *Growth of* Al $(11\bar{2}0)$ *on* 6H-SiC $(11\bar{2}0)$ *by Molecular-Beam Epitaxy*, Jpn. J. Appl. Phys. **41**, L1348 (2002)

[178] T. Onuma, H. Amaike, M. Kubota, K. Okamoto, H. Ohta, J. Ichihara, H. Takasu and S. F. Chichibu: *Quantum-confined Stark effects in the m-plane* $In_{0.15}Ga_{0.85}N$ *multiple quantum well blue light-emitting diode fabricated on low defect density freestanding* GaN *substrate*, Appl. Phys. Lett. **91**, 181903 (2007)

[179] M. C. Payne, M. P. Teter, D. C. Allan, T. A. Arias and J. D. Joannopoulos: *Iterative minimization techniques for ab initio total-energy calculations: molecular dynamics and conjugate gradients*, Rev. Mod. Phys. **64**, 1045 (1992)

[180] X. Peng, M. C. Schlamp, A. V. Kadavanich and A. P. Alivisatos: *Epitaxial Growth of Highly Luminescent CdSe/CdS Core/Shell Nanocrystals with Photostability and Electronic Accessibility*, J. Am. Chem. Soc. **119**, 7019 (1997)

[181] J. P. Perdew, K. Burke and M. Ernzerhof: *Generalized Gradient Approximation Made Simple*, Phys. Rev. Lett. **77**, 3865 (1996)

[182] A. Petersson, A. Gustafsson, L. Samuelson, S. Tanaka and Y. Aoyagi: *Cathodoluminescence spectroscopy and imaging of individual* GaN *dots*, Appl. Phys. Lett. **74**, 3513 (1999)

[183] M. Petrov, L. Lymperakis, J. Neugebauer, R. Stefaniuk and P. Dłużewski: *Nonlinear elastic effects in group III-Nitrides: From ab initio to finite element calculation*, Proc. 17th Int. Conf. on Computer Methods in Mechanics, Spała (2007)

[184] S. I. Petrov, A. P. Kaĭdash, D. M. Krasovitskiĭ, I. A. Sokolov, Y. V. Pogorel'skiĭ, V. P. Chalyĭ, A. P. Shkurko, M. V. Stepanov, M. V. Pavlenko and D. A. Baranov: *MBE of InGaN/GaN Heterostructures using Ammonia as a Source of Nitrogen*, Tech. Phys. Lett. **30**, 580 (2004)

[185] J. C. Phillips: *Bonds and bands in semiconductors* (Academic Press, New York, 1973)

[186] G. E. Pikus and G. L. Bir: *Effect of deformation on the hole energy spectrum of germanium and silicon*, Fiz. Tverd. Tela (Leningrad) **1**, 1642 (1959) [Sov. Phys. Solid State **1**, 1502 (1959)]

[187] F. H. Pollak and M. Cardona: *Piezo-Electroreflectance in* Ge, GaAs *and* Si, Phys. Rev. **172**, 816 (1967)

[188] F. A. Ponce and D. P. Bour: *Nitride-based semiconductors for blue and green light-emitting devices*, Nature (London) **386**, 351 (1997)

[189] M. Posternak, A. Baldereschi, A. Catellani and R. Resta: *Ab initio study of the Spontaneous Polarisation of Pyroelectric* BeO, Phys. Rev. Lett. **64**, 1777 (1990)

[190] M. Povolotskyi, M. Auf der Maur and A. Di Carlo: *Strain effects in freestanding three-dimensional nitride nanostructures*, Phys. Stat. Solidi (c) **2**, 3891 (2005)

[191] C. Pryor, M. E. Pistol and L. Samuelson: *Electronic structure of strained $InP/Ga_{0.51}In_{0.49}P$ quantum dots*, Phys. Rev. B **56**, 10404 (1997)

[192] C. Pryor: *eight band calculations of strained InAs/GaAs quantum dots compared with one-, four-, and six-band approximations*, Phys. Rev. B **57**, 7190 (1998)

[193] F. Qian, Y. Li, S. Gradečak, D. Wang, C. J. Barrelet and C. M. Lieber: *Gallium Nitride-Based Nanowire Radial Heterostructures for Nanooptics*, Nano Lett. **4**, 1975 (2004)

[194] S. B. Radhia, K. Boujdaria, S. Ridene, H. Bouchriha and G. Fishman: *Band structures of GaAs, InAs, and Ge: A 24-$\mathbf{k}\cdot\mathbf{p}$ model*, J. Appl. Phys. **94**, 5726 (2003)

[195] M. Razeghi and A. Rogalski: *Semiconductor ultraviolet detectors*, J. Appl. Phys. **79**, 7433 (1996)

[196] J. P. Reithmaier and A. Forchel:*Semiconductor quantum dots*, IEEE Circuits and Devices Magazine **19**, 24 (2003)

[197] R. Resta: *Macroscopic polarisation in crystalline dielectrics: the geometric phase approach*, Rev. Mod. Phys. **66**, 899 (1994)

[198] D. H. Rich, H. T. Lin and A. Larsson: *Influence of defects on electron-hole plasma recombination and transport in a nipi-doped $In_xGa_{1-x}As/GaAs$ multiple quantum well structure*, J. Appl. Phys. **77**, 12 (1995)

[199] S. Richard, F. Aniel and G. Fishman: *Energy-band structure of Ge, Si, and GaAs: A thirty-band $\mathbf{k}\cdot\mathbf{p}$ method*, Phys. Rev. B **70**, 235204 (2004)

[200] S. Richard, F. Aniel and G. Fishman: *Band diagrams of Si and Ge quantum wells via the 30-band $\mathbf{k}\cdot\mathbf{p}$ method*, Phys. Rev. B **72**, 245316 (2005)

[201] P. Rinke, A. Qteish, J. Neugebauer, C. Freysoldt and M. Scheffler: *Combining GW calculations with exact-exchange density-functional theory: an analysis of valence-band photoemission for compound semiconductors*, New. J. Phys. **7**, 126 (2005)

[202] P. Rinke, M. Winkelnkemper, A. Qteish, D. Bimberg, J. Neugebauer and M. Scheffler: *Consistent set of band parameters for the group-III nitrides AlN, GaN, and InN*, Phys. Rev. B **77**, 075202 (2008)

[203] F. Rol, S. Founta, H. Mariette, B. Daudin, L. S. Dang, J. Bleuse, D. Peyrade, J. M. Gérard and B. Gayral: *Probing exciton localisation in nonpolar GaN/AlN quantum dots by single-dot optical spectroscopy*, Phys. Rev. B **75**, 125306 (2007)

[204] S. J. Rosner, E. C. Carr, M. J. Ludowise, G. Girolami and H. I. Erikson: *Correlation of cathodoluminescence inhomogeneity with microstructural defects in epitaxial GaN grown by metalorganic chemical-vapor deposition*, Appl. Phys. Lett. **70**, 420 (1997)

[205] F. Rossi: *The excitonic quantum computer*, IEEE Trans. Nanotechnology **3**, 165 (2004)

[206] J. L. Rouvière, J. Simon, N. Pelekanos, B. Daudin and G. Feuillet: *Preferential nucleation of GaN quantum dots at the edge of AlN threading dislocations*, Appl. Phys. Lett. **75**, 2632 (1999)

[207] C. Santori, M. Pelton, G. Solomon, Y. Dale and Y. Yamamoto: *Triggered Single Photons from a Quantum Dot*, Phys. Rev. Lett. **86**, 1502 (2001)

[208] A. Schliwa, M. Winkelnkemper and D. Bimberg: *Few-particle energies versus geometry and composition of $In_xGa_{1-x}As/GaAs$ self-organized quantum dots*, Phys. Rev. B **79**, 075443 (2009)

[209] H. C. Schneider, W. W. Chow and S. W. Koch: *Many-body effects in the gain spectra of highly excited quantum-dot lasers*, Phys. Rev. B **64**, 115315 (2001)

[210] S. Schulz and G. Czycholl: *Tight-binding model for semiconductor nanostructures*, Phys. Rev. B **72**, 165317 (2005)

[211] S. Schulz, S. Schumacher and G. Czycholl: *Tight-binding model for semiconductor quantum dots with a wurtzite crystal structure: From one-particle properties to Coulomb correlations and optical spectra*, Phys. Rev. B **73**, 245327 (2006)

[212] S. Schulz, A. Berube and E. P. O'Reilly: *Polarisation fields in nitride-based quantum dots grown on nonpolar substrates*, Phys. Rev. B **79**, R081401 (2009)

[213] P. Sharma and X. Zhang: *Impact of size-dependent non-local elastic strain on the electronic band structure of embedded quantum dots*, Proc. IMechE **220** Part N (2006)

[214] J. J. Shi and Z. Z. Gan: *Effects of piezoelectricity and spontaneous polarization on localized excitons in self-formed InGaN quantum dots*, J. Appl. Phys. **94**, 407 (2003)

[215] S. Simhony, E. Kapon, E. Colas, D. M. Hwang, N. G. Stoffel and P. Worland: *Vertically stacked multiple-quantum-wire semiconductor diode lasers*, Appl. Phys. Lett. **59**, 2225 (1991)

[216] J. Simon, N. T. Pelekanos, C. Adelmann, E. Martinez-Guerrero, R. Andre, B. Daudin, L. S. Dang and H. Mariette: *Direct comparison of recombination dynamics in cubic and hexagonal GaN/AlN quantum dots*, Phys. Rev. B **68**, 035312 (2003)

[217] I. Souza, N. Marzari and D. Vanderbilt: *Maximally localized Wannier functions for entangled energy bands*, Phys. Rev. B **65**, 035109 (2001)

[218] J. S. Speck and S. J. Rosner: *The role of threading dislocations in the physical properties of GaN and its alloys*, Physica B **273-274**, 24 (1999)

[219] *www.sphinxlib.de*

[220] J. Stangl, V. Holý and G. Bauer: *Structural properties of self-organized semiconductor nanostructures*, Rev. Mod. Phys. **76**, 725 (2004)

[221] J. Stanley and N. Goldsman: *Full-zone hole dispersion relations in Si using Schur-complement decomposition*, Phys. Rev. B **51**, 4931 (1995)

[222] J. Stark: *Beobachtungen über den Effekt des elektrischen Feldes auf Spektrallinien I. Quereffekt*, Ann. Phys. **43**, 965 (1914)

[223] O. Stier: *Electronic and optical properties of quantum dots and wires*, PhD thesis, Berlin (2000)

[224] O. Stier, M. Grundmann and D. Bimberg: *Electronic and optical properties of strained quantum dots modeled by 8-band* $\mathbf{k} \cdot \mathbf{p}$ *theory*, Phys. Rev. B **59**, 5688 (1999)

[225] I. N. Stranski and V. L. Krastanow, Akad. Wiss. Lit. Mainz Abh. Math. Naturwiss. Kl. **146**, 797 (1939)

[226] Y. J. Sun: *Growth and characterization of M-plane GaN and InGaN/GaN multiple quantum wells*, PhD thesis, Berlin (2003)

[227] K. Suzuki, R. A. Hogg and Y. Arakawa: *Structural and optical properties of type II GaSb/GaAs self-assembled quantum dots grown by molecular beam epitaxy*, J. Appl. Phys. **85**, 8349 (1999)

[228] K. Tachibana, T. Someya and Y. Arakawa: *Nanometer-scale InGaN self-assembled quantum dots grown by metalorganic chemical vapor deposition*, Appl. Phys. Lett. **74**, 383 (1999)

[229] T. Takano, Y. Narita, A. Horiuchi and H. Kawanishi: *Room-temperature deep-ultraviolet lasing at 241.5 nm of AlGaN multiple-quantum-well laser*, Appl. Phys. Lett. **84**, 3567 (2004)

[230] R. N. Thurston and K. Brugger: *Third-Order Elastic Constants and the Velocity of Small Amplitude Elastic Waves in Homogeneously Stressed Media*, Phys. Rev. **133**, A1604 (1964)

[231] A. T. Tilke, F. C. Simmel, H. Lorenz, R. H. Blick and J. P. Kotthaus: *Quantum interference in a one-dimensional silicon nanowire*, Phys. Rev. B **68**, 075311 (2003)

[232] K. Tsubouchi and N. Mikoshiba: *Zero-Temperature-Coefficient SAW Devices on AlN Epitaxial-Films*, IEEE Trans. Sonics Ultrason. **32**, 634 (1985)

[233] V. M. Ustinov, N. A. Maleev, A. E. Zhukov, A. R. Kovsh, A. Yu. Egorov, A. V. Lunev, B. V. Volovik, I. L. Krestnikov, Yu. G. Musikhin, N. A. Bert, P. S. Kop'ev, and Z. I. Alferov: InAs/InGaAs *quantum dot structures on* GaAs *substrates emitting at* 1.3 μm, Appl. Phys. Lett. **74**, 2815 (1999)

[234] C. G. Van de Walle: *Band lineups and deformation potentials in the model solid theory*, Phys. Rev. B **39**, 1871 (1989)

[235] G. Vaschenko, D. Patel, C. S. Menoni, N. F. Gardner, J. Sun, W. Götz, C. N. Tomé and B. Clausen: *Significant strain dependence of piezoelectric constants in* $In_xGa_{1-x}N/GaN$ *quantum wells*, Phys. Rev. **B** 64, 241308(R) (2001)

[236] M. Vollmer and A. Weber: *Keimbildung in übersättigten Gebilden*, Z. Phys. Chem. **119**, 277 (1926)

[237] L. C. Lew Yan Voon and M. Willatzen: *The* **k · p** *Method* (Springer, Berlin, 2009)

[238] I. Vurgaftman, J. R. Meyer and L. R. Ram-Mohan: *Band parameters for III-V compound semiconductors and their alloys*, J. Appl. Phys. **89**, 5815 (2001)

[239] I. Vurgaftman and J. R. Meyer: *Band parameters for nitrogen-containing semiconductors*, J. Appl. Phys. **94**, 3675 (2003)

[240] M. Wahn and J. Neugebauer: *Generalized Wannier functions: An efficient way to construct ab-initio tight-binding parameters for group-III nitrides*, Phys. Stat. Solidi (b) **243**, 1583 (2006)

[241] P. Waltereit, O. Brandt, A. Trampert, H. T. Grahn, J. Menninger, M. Ramsteiner, M. Reiche and K. H. Ploog: *Nitride semiconductors free of electrostatic fields for efficient white light-emitting diodes*, Nature (London) **406**, 865 (2000)

[242] M. Walther, E. Kapon, E. Colas, D. M. Hwang and R. Bhat: *Carrier capture and quantum confinement in GaAs/AlGaAs quantum wire lasers grown on V-grooved substrates*, Appl. Phys. Lett. **60**, 521 (1991)

[243] J. Wang, M. S. Gudiksen, X. Duan, Y. Cui and C. M. Lieber: *Highly Polarized Photoluminescence and Photodetection from Single Indium Phosphide Nanowires*, Science **293**, 1457 (2001)

[244] L. W. Wang and A. Zunger: *Local-density-derived semiempirical pseudopotentials*, Phys. Rev. B **51**, 17398 (1995)

[245] L. W. Wang, J. Kim and A. Zunger: *Electronic structures of [110]-faceted self-assembled pyramidal InAs/GaAs quantum dots*, Phys. Rev. B **59**, 5678 (1999)

[246] X. Weng, R. A. Burke and J. M. Redwing: *The nature of catalyst particles and growth mechanisms of* GaN *nanowires grown by* Ni-*assisted metal-organic chemical vapor deposition*, Nanotechnology **20**, 085610 (2009)

[247] F. Widmann, B. Daudin, G. Feuillet, Y. Samson, M. Arlery and J. L. Rouvière: *Evidence of 2D-3D transition during the first stages of* GaN *growth on* AlN, MRS. Internet J. Nitride Semicond. Res. **2**, 20 (1997)

[248] F. Widmann, B. Daudin, G. Feuillet, Y. Samson, J. L. Rouvière and N. T. Pelekanos: *Growth kinetics and optical properties of self-organized* GaN *quantum dots*, J. Appl. Phys. **83**, 7618 (1998)

[249] F. Widmann, J. Simon, B. Daudin, G. Feuillet, J. L. Rouvière, N. T. Pelekanos and G. Fishman: *Blue-light emission from* GaN *self-assembled quantum dots due to giant piezoelectric effect*, Phys. Rev. B **58**, R15989 (1998)

[250] D. P. Williams, A. D. Andreev and E. P. O'Reilly: *Dependence of exciton energy on dot size in* GaN/AlN *quantum dots*, Phys. Rev. B **73**, R241301 (2006)

[251] M. Winkelnkemper, A. Schliwa and D. Bimberg: *Interrelation of structural and electronic properties in* $In_xGa_{1-x}N$/GaN *quantum dots using an eight band* $\mathbf{k} \cdot \mathbf{p}$ *model*, Phys. Rev. B **74**, 155322 (2006)

[252] A. Wojs, P. Hawrylak, S. Fafard and L. Jacak: *Electronic structure and magneto-optics of self-assembled quantum dots*, Phys. Rev. B **54**, 5604 (1996)

[253] T. Yamamoto, M. Kasu, S. Noda and A. Sasaki: *Photoluminescent properties and optical absorption of* AlAs/GaAs *disordered superlattices*, J. Appl. Phys. **68**, 15 (1990)

[254] H. Yang, L. X. Zheng, J. B. Li, X. J. Wang, D. P. Xu, Y. T. Wang, X. W. Hu and P. D. Han: *Cubic-phase* GaN *light-emitting diodes*, Appl. Phys. Lett. **74**, 2498 (1999)

[255] T. D. Young and O. Marquardt: *Influence of strain and polarisation on electronic properties of a* GaN/AlN *quantum dot*, Phys. Stat. Solidi (c) **6**, 557 (2009)

[256] P. Y. Yu, M. Cardona: *Fundamentals of Semiconductors* (Springer, 1996)

[257] P. Zeeman: *On the influence of Magnetism on the Nature of the Light emitted by a Substance*, Phil. Mag. **43**, 226 (1897)

[258] L. Zhou, J. E. Epler, M. R. Krames, W. Goetz, M. Gherasimova, Z. Ren, J. Han, M. Kneissl, and N. M. Johnson: *Vertical injection thin-film* AlGaN/AlGaN *multiple-quantum-well deep ultraviolet light-emitting diodes*, Appl. Phys. Lett. **89**, 241113 (2006)

[259] M. Zieliński, W. Jaskólski, J. Aizpurua and G. W. Bryant: *Strain and spin-orbit effects in self-assembled quantum dots*, Acta Phys. Pol. A **108**, 929 (2005)

[260] O. Zitouni, K. Boujdaria and H. Bouchriha: *Band parameters for GaAs and Si in the 24-**k** · **p** model*, Semicond. Sci. Technol. **20**, 908 (2005)

[261] A. Zrenner, E. Beham, S. Stufler, F. Findeis, M. Bichler and G. Abstreiter: *Coherent properties of a two-level system based on a quantum-dot photodiode,* Nature **418**, 612 (2002)

[262] A. Zunger: *Electronic-Structure Theory of Semiconductor Quantum Dots*, MRS Bulletin, p. 35-42 (1998)

[263] A. Zunger: *Pseudopotential Theory for Semiconductor Quantum Dots*, Phys. Stat. Solidi (b) **224**, 727 (2001)

I want morebooks!

Buy your books fast and straightforward online - at one of world's fastest growing online book stores! Environmentally sound due to Print-on-Demand technologies.

Buy your books online at
www.morebooks.shop

Kaufen Sie Ihre Bücher schnell und unkompliziert online – auf einer der am schnellsten wachsenden Buchhandelsplattformen weltweit! Dank Print-On-Demand umwelt- und ressourcenschonend produziert.

Bücher schneller online kaufen
www.morebooks.shop

KS OmniScriptum Publishing
Brivibas gatve 197
LV-1039 Riga, Latvia
Telefax:+371 686 204 55

info@omniscriptum.com
www.omniscriptum.com

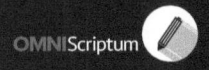

Printed by Books on Demand GmbH, Norderstedt / Germany